Regnaldo Gouveia dos Santos

Socio-environmental impacts on the banks of the São Francisco River

Regnaldo Gouveia dos Santos

Socio-environmental impacts on the banks of the São Francisco River

The relationship between man and nature

ScienciaScripts

Imprint

Cover image: www.ingimage.com

This book is a translation from the original published under ISBN 978-3-330-76462-0.

Publisher:
Sciencia Scripts
is a trademark of
Dodo Books Indian Ocean Ltd. and OmniScriptum S.R.L publishing group

120 High Road, East Finchley, London, N2 9ED, United Kingdom
Str. Armeneasca 28/1, office 1, Chisinau MD-2012, Republic of Moldova, Europe
Managing Directors: Ieva Konstantinova, Victoria Ursu
info@omniscriptum.com

Printed at: see last page
ISBN: 978-620-8-50478-6

I dedicate this work to all the exhibitors at the Modern Art Week (1922), which, among other factors, instigated me to publish this book among many others that I hope will be published. I would like to dedicate it mainly to those who were not featured in the media at the time, because we know about the great impasses of this medium to this day, despite the existence of a Federal Law in favour of national culture.

ACKNOWLEDGEMENTS

Above everything and all things, to God (Holy Trinity).

To Professor Dr Neli A. Mello Théry, for her guidance in the development of the study that was important for the existence of this work.

To all the interviewees, without exception, from representatives of public organisations to people with lower levels of education and those who had no level of education, but were well informed about the subject of the interviews, who collaborated in providing them.

To my wife, my son Dêjeces and my parents.

To all the professors who contributed to my knowledge, such as Dr José Conti, Dr Hérve Théry, Dr Celso Clemente, Dr Marcelo Martineli, among others.

To Prof Dr Hunaldo Lima, for his significant contributions.

To the memorable Professor Maria Elza Aragão.

To all those who directly and indirectly contributed to the realisation of this study.

CONTENTS

CHAPTER 1

GEO-HISTORY AND THE NATURE OF PLACE

This chapter aims to show the origins of Porto da Folha/SE as a municipality and as a region. As the town is located in the São Francisco River basin and, above all, on its banks, and has developed in its favour, the history of this river was covered, as it is the main river in this basin. At the same time, the Capivara River was researched because of its influence on the municipality of Porto Folhense, whose centre is bathed by this tributary of the São Francisco. For this reason, an approach is also made to characterising the nature of the place, as it was during the rice cycle, due to the natural phenomenon that occurred through the São Francisco River and reached the Capivara, which bathes the main local floodplain, Ilha do Ouro.

It was possible to contextualise the geo-history and nature of the environment of Porto da Folha/SE using this method, which is considered relevant to understanding the subject dealt with throughout the work. It is from this analysis, from the comparison of the present with the past, which makes it easier to understand the new spatial configurations of the place.

A GEOGRAPHICAL VIEW OF THE PLACE COVERING THE GEO-HISTORY OF THE SÃO FRANCISCO RIVER AND TWO TRIBUTARIES ON THE SERGIPANO SIDE

The São Francisco River rises in the Serra da Canastra, Minas Gerais, passes through the states of Bahia and Pernambuco and flows into the Atlantic Ocean between the states of Alagoas and Sergipe, reaching a length of around 3,000 kilometres.

Figure 1. Source of the São Francisco River, Minas Gerais. Source: photo by the author, July 2008 .[1]

Although there are several points on the Serra da Canastra plateau where water rises in rainy weather, indicating the source of the river, figure (01) shows the main place where the water of the "Velho Chico" (Old Chico), as it is also affectionately called by the local people, rises. This fact was better recorded at the time of the photograph because it was a dry period in the region.

At the top of the picture (no. 1), on the left, is the sign pointing in the direction of the arrow indicating the source of the river. In the lower part of the picture, you can see the undergrowth of the savannah and a good part of the extensive plateau of the Serra da Canastra. In the background is the watershed that determines the domain of the São Francisco spring. In the centre is a statue of São Francisco, whose name gave rise to the river. The name was given by European navigators in 1501, when they first caught sight of the flowing river.

Porto da Folha/SE is located between the geographical coordinates of 09° 55' South latitude and 37° 17' West longitude. It is bordered by the Sergipe municipalities of Gararu, Monte Alegre de Sergipe, Poço Redondo, Nossa Senhora da Glória and the state of Alagoas by the São Francisco River.

The town's municipal centre is on the plateau between the Serra da Lagoa Comprida (or Serra de Campinho), Serra da Cal and Serra dos Homens and is

[1] Authorised by the Ministry of the Environment (MMA) - Chico Mendes Institute for Biodiversity Conservation (Icmcb), July 2008.

around 180 km from the capital of Sergipe, if you travel on paved roads. The city's altitude exceeds 50 metres and does not reach 100 metres, i.e. these measurements are in relation to sea level. In 2010, according to Ibge (2010), the municipality had a population of 26,636. The Xingó Hydroelectric Plant is about 80 kilometres upstream (in the direction of the river's source) from Porto da Folha/SE and is located on the São Francisco River, in the municipality of Canindé de São Francisco/SE, as can be seen on the map (fig. 2) below.

The Ilha do Ouro floodplain covers around 260 hectares, although the whole of the extensive river plain bordering the Capivara River, from the edge of the São Francisco River, encompasses the vicinity of the municipal centre of Porto Folhense. This is a place where its size extends even further, meeting Lagoa Comprida and Lagoa de Miguel Pinto (popular names). The latter is on the banks of the River Capivara and closer to the urban area than the other lagoon. Thus, the entire length of the river plain can exceed 10 (ten) kilometres.

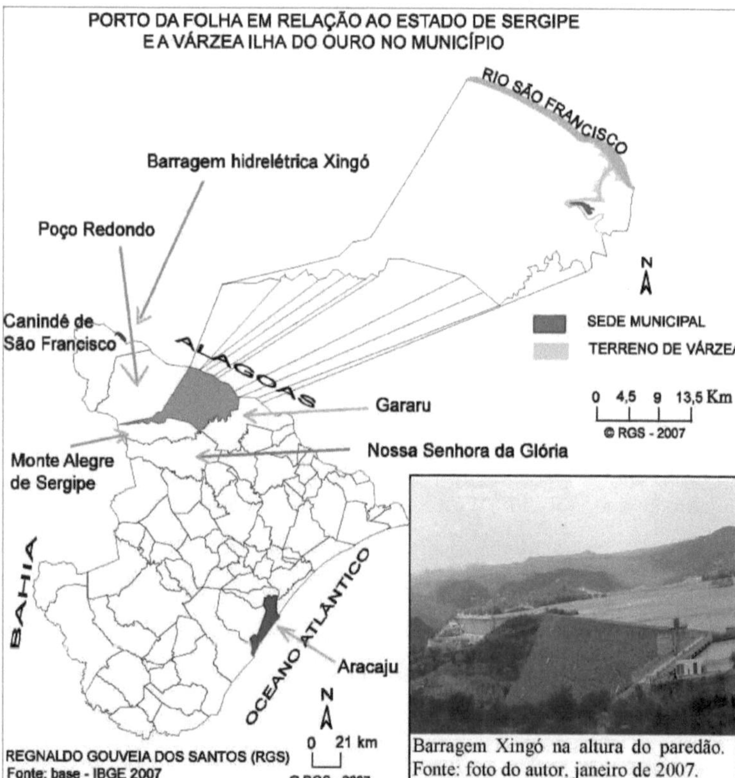

Barragem Xingó na altura do paredão.
Fonte: foto do autor, janeiro de 2007.

Figure 2 - Location map of the municipality of Porto da Folha/SE and its main floodplain.

Analysing the map (fig. 2), it is possible to see Porto da Folha/SE in relation to several of the points mentioned above, as well as the fact that the main floodplain is to the north-east of the municipal centre. Below, on the right-hand side of the picture, you can see part of the lake of the Xingó Hydroelectric Power Station, with an emphasis on its wall crossing the São Francisco River from the territory of Sergipe, on the left-hand side of the picture, to Alagoas, on the right-hand side.

Historically, it is known that when Pedro Álvares Cabral discovered the island of Vera Cruz (Brazil) in 1500, around a year later Américo Vespúcio reached the mouth of the São Francisco River via the Atlantic Ocean. Until then, the river was called "Opará" by local primitives, which would be something like "river-sea". Due to the fact that the navigator made this discovery from the perspective of European civilisation and, above all, the fact that his calendar marked October 4, 1501, the day of Saint Francis, from then on this was the name that European navigators gave to the river.

Later, the São Francisco River became one of the main means of access for exploring the interior of Brazilian lands, as several cattle corrals were set up on its banks, which contributed to regional territorial expansion. It was in the middle of the 16th century that considerable tracts of land began to be explored, contributing to population growth in the regions where the corrals were installed. This is why the river also came to be known as "Rio-dos-Currais" (River of the Corrals), but the initial name given by the first crew to reach its mouth persists to this day.

This river has its source about 2,900 kilometres from the area studied here, and if you consider its course from its source to its mouth (the Atlantic Ocean), it is 3,000 kilometres long.

Figure 3 below shows the first high-level waterfall of the São Francisco River, inside the Serra da Canastra National Park, where its source is located.

In the upper part of this picture (no. 3), you can see the crack caused by nature's physicochemical effect on the immense rock that forms this mountain range. On the sides of the waterfall, you can see cracks starting at the foot of the mountain, located close to the water. From the top, where the water flows down to reach the natural pool far below, you can see the Casca D'anta waterfall, which is around 180 metres high. This waterfall caught the attention of ancient scholars such as Mesquita (192-?).

Figure 3: Casca D'anta waterfall, Serra da Canastra National Park - MG. Source: photo by the author (authorised by the Ministry of the Environment (MMA) - Chico Mendes Institute for Biodiversity Conservation (Icmcb)), July 2008.

The map (fig. 4) below shows that, after travelling through the state of Minas Gerais, from the Serra da Canastra, the river continues through the state of Bahia, bordering part of Pernambuco, and flows into the Atlantic Ocean, between Sergipe and Alagoas, making a natural border. The São Francisco river basin, covering some 640,000 kilometres[2], also encompasses part of Goiás and the Federal District, with the source states of Goiás and Bahia bathed by the river.

for the largest percentage (more than 80 per cent) of the portion, among all the states that make up this area.

The great longitudinal length of the São Francisco River is divided into four fragments: Upper, Middle, Middle and Lower reaches. The latter runs from Paulo Afonso/BA to the mouth. The Middle São Francisco comprises the area between that city and Remanso/BA. The Middle São Francisco is located between this city in Bahia and the mouth of the Jequitaí River, in the state of Minas Gerais. The upper course of the river is located in the central part of the state of Minas Gerais, between the headwaters where the source of the São Francisco is located and the mouth of the Jequitaí River. The Upper São Francisco is at an altitude that varies from 1,200 metres to 470 metres from the headwaters to the point where the Jequitaí meets the São Francisco River.

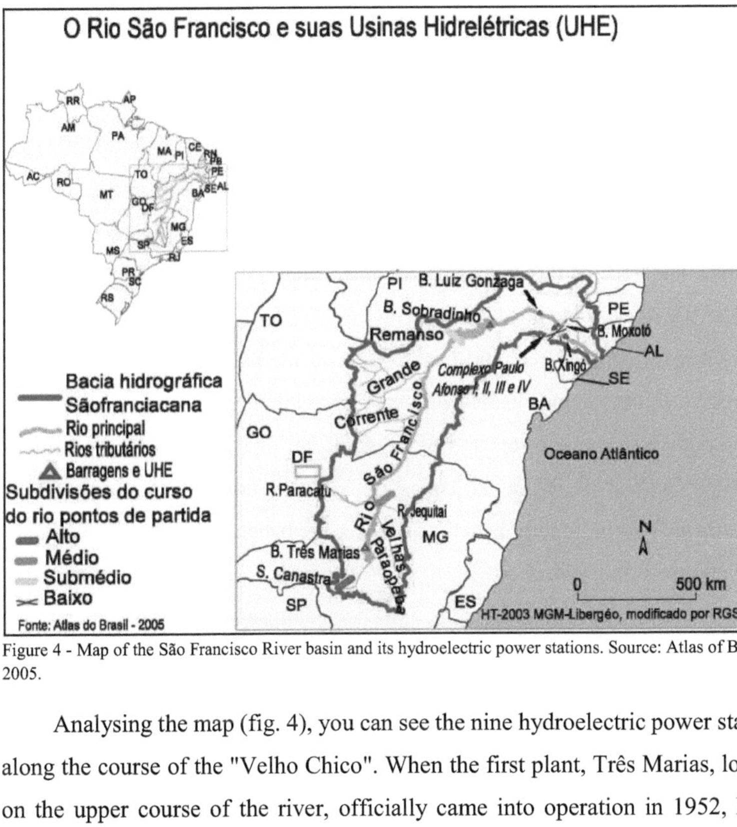

Figure 4 - Map of the São Francisco River basin and its hydroelectric power stations. Source: Atlas of Brazil - 2005.

Analysing the map (fig. 4), you can see the nine hydroelectric power stations along the course of the "Velho Chico". When the first plant, Três Marias, located on the upper course of the river, officially came into operation in 1952, Paulo Afonso (PA) I, on the lower course, had already begun construction between 1948 and 1949, but was only inaugurated in 1955. PA II came into operation in 1961 and the last two hydroelectric power stations in the Paulo Afonso complex, PA III and PA IV, began operating in the 1970s, with the first of these dating back to 1971.

In 1977, the Moxotó plant, which is located at the same point on the river as the Paulo Afonso hydroelectric complex, started operating with its first generating machine. The Sobradinho hydroelectric power station began generating energy in 1979, although its dam was built in 1973. It is also located in the São Francisco sub-medium, where the Luiz Gonzaga (formerly Itaparica) power station is located, and began operating in 1988. The Xingó Hydroelectric Power Station is located in the Lower São Francisco and was the last of the power stations to start operating, in 1994, with its first generating machine.

Geographically, socio-spatial expansion on the banks of rivers often led to the territorial organisation of the surrounding area, as was the case with the exploitation of the São Francisco river basin.

One of the most important factors for expansion was the availability of water. Because of this, in the 16th century, the main means of transport in this territory was by river, which was the most efficient means of travelling when it came to exploring the lands of this river area. In other words, this space is precisely the scope of the river, subject to human action on nature, an example of which was man's takeover of its waters and banks. Later, cattle and horse breeding gained importance in the region, increasing social development and the relations that existed between the colonisers and the Royal Court of Portugal, mediated by the Governor General of Bahia de Todos os Santos. Thus, demographic growth gradually took place, mainly with the arrival of the bandeiras and, later, due to the exploitation of the sesmarias. From 1590 onwards, according to Woortmann (1995), spatial occupation began to be linked to colonisers from Bahia, such as the powerful Garcia D'Ávila, and Pernambuco.

As Ferreira (1959) rightly observed, the old Curral do Buraco, which gave rise to the municipal centre of Porto da Folha/SE, was upstream from the Curral de Pedras. These were among the various corrals installed on the banks of the São Francisco River.

Nowadays, the statement that Curral do Buraco was one of the corrals of the famous Rio-dos-Currais may come as a surprise to those unfamiliar with the region's geography and history. This surprise is due to the significant transformation that the local area has undergone as a result of man's action on nature, especially after the construction of the Xingó Dam. As a result, the small town that used to be bathed by the waters of the São Francisco due to the large volumes of cyclical flooding over the extensive river plain is now about 6 kilometres away from the São Francisco water mirror due to the extinction of the river's flood cycle, as explained in Chapter 3.

As has already been said, if you turn to geography, you'll soon realise that this place is bathed by the Capivara River, a tributary of the São Francisco. The latter, in turn, until the end of the 20th century, was a determining factor in the volume of this tributary, making it perennial until it turned around in the district of the city of Porto da Folha/SE, which is no longer possible to observe today. The above assertion is confirmed because, in relation to rice transport in the 20th century, "in those days there were no lorries, it was transported by canoe [...]. There was, there was the river, the canoe would pick up here at... it would come to the rocks, sometimes it would be full, it would pick up here [at Rua de Cima]. [...] Porto das Pedra, it's in Miguel Chorão, it's[2] "! It was also noted that Porto da Folha/SE had a commercial relationship with the riverside towns of Propriá/SE and Penedo/AL, where the region's most important fairs were held. Porto da Folha's population used the River Capivara to travel to these towns from the municipal centre.

The famous Curral was located in the district of Tanque Novo, which is now urbanised, and is in the southeast of the city in relation to the Mother Church. It is possible to see that the Curral was in a strategic location, as the tank was huge and at a certain altitude in the rugged local terrain, so it was protected from the floods of the São Francisco that reached the district via the Capivara. Even so, the Curral was in the vicinity of this creek, which was important for the development of rural activities.

The Capivara River is one of the main tributaries of the São Francisco in Sergipe. It is classified as an intermittent river due to the absence of water at some points along its course during the dry season. For this reason, it is usually only possible to observe normality, an abundance of water, in its bed, in the district of the municipal centre, if its natural floods occur upstream and downstream. Floods from the last direction of the river are restricted to the torrential floods of the São Francisco, i.e. when this phenomenon occurred or as long as it exists, even if it is

[2] The source has not been revealed for reasons of preservation of rights.

acyclic. According to Santos and Andrade (1992), the Capivara rises between the pediplano sertanejo and the plateau of the south-western Serra Negra, which forms the natural border between Sergipe and Bahia.

Analysing the map (fig. 05) below, you can see that from its source until it flows into the São Francisco, this tributary receives several intermittent rivers (or streams), with the Cachorro standing out as one of the most important. This stream rises in the same region, a little to the north of the source of the Capivara. In times of flooding, the two flow simultaneously and in parallel, coming together in the municipality of Monte Alegre de Sergipe.

Figure 5. The Rio or Riacho Capivara and its main tributaries, as well as the Rio Gararu/SE.

From the aforementioned municipality, the main riverbed continues downstream (in the direction of the river's mouth) to its confluence with the Riacho da Mão Esquerda, from where it forms a natural border between the municipalities of Gararu/SE and Porto da Folha/SE until it leaves in the direction of the seat of the latter municipality. After reaching the plains of the Porto da Folha floodplains, the Capivara flows into the São Francisco River.

On the map (fig. 5) above, you can also see that the Rio Capivara starts from its source under the name Rio do Sal.

According to the cartography of Sergipe Del Rei (some dates cannot be specified because they are very old works), the current Rio or Riacho Capivara was once called Rio Ilha do Ouro and Rio da Porteira. This dates back to the beginning of the Captaincy of Sergipe (mid-16th century). At the same time as the second river name appears, there was also the name of Ilha do Ouro, the present-day settlement, which may point to the origin of the farm with this name.

In a letter from the early days (APPENDIX A), it can be seen that the name Porto da Folha referred to the entire scope of a large territory on the banks of the São Francisco River. It can be seen that this entire region was the Sesmaria of Burgos and others, a link to the Gomes Castello Branco family, with the neighbouring Sesmaria of Garcia D'Avila, who was one of the pioneers of territorial occupation in Sergipe, coming from Bahia de Todos os Santos.

Analysing the map of the state of Sergipe and part of the state of Bahia (APPENDIX B) confirms the above assertions, noting that the name Ilha do Ouro stands out as a point on the edge of the São Francisco River, where the farm of that name was located. Despite this, it can be seen on this map that the attribute represented by the Ilha do Ouro River, possibly the Capivara, was not linked to this point. If you look at the current map, you can see that it flows exactly upstream (towards the source of the river) from the village, in relation to the São Francisco River.

The letter from the Comarca de Sergipe Del Rei [1807 or 1809] (APPENDIX C) points out that Porto da Folha was on the banks of the River São Francisco, as was Curral de Pedras, now Gararu/SE, and that Porto da Folha was upstream from Gararu. However, Porto da Folha was far from the Rio da Porteira, today's Capivara, which flows into the Sergipe side, opposite the present-day Rio Barra do Ipanema on the Alagoas side.

According to Souza (1944), Vila Nova was home to riches from the mineral kingdom, as well as gold. The places indicated in the aforementioned cartography belonged to this town at the time. Thus, it is possible that the name Ilha do Ouro originated because the existence of gold was pointed out in its direction. Or rather, it was pointed out in what is now the Capivara Creek, which, next to the village, flows into the São Francisco River. In reference to this, in "[...] Capivara was found - a gold leaflet that had 6 octaves." (SOUZA, 1944, p. 41).

When looking at all the cartography mentioned above, no names were found for Capivara. However, it can be seen that the contemporary cartography of Sergipe (see figure 5) mentions Capivara as the watercourse that surrounds the municipal centre of Porto da Folha/SE and flows a few steps upstream from the village of Ilha do Ouro, in relation to the São Francisco. You can also see that part of the town's name is related to the expression gold leaf.

Ilha do Ouro had long been the main harbour that served Porto da Folha/SE when it was a town, so that's where the produce that arrived there flowed, usually via the Capivara.

As for the fact observed on the map of the State of Sergipe and part of the State of Bahia (APPENDIX B), where the Ilha do Ouro River flows into the São Francisco, but far from the point (now Lugarejo) that had the same name as that river, this could already be the origin of the old Ilha do Ouro Farm. According to this study, there are two possible explanations for this.

The first is that the farm's headquarters could be located somewhere other than where it is today. Today, the main house is still standing in Povoado Ilha do Ouro. However, its structure does not appear to have lasted for more than four hundred years, according to the documentation analysed (taking into account the beginning of the exploration of the São Francisco). Thus, the farm would have been located on the opposite side of the river from what was seen in the 20th century.

The second explanation is based on the following premise: the river that, according to the map of the State of Sergipe and part of the State of Bahia (APPENDIX B), is the Porto da Folha as it is represented on that map, cannot be born in the Serra Negra. As a result, the old map of the Sergipe Del Rei District [1807 or 1809] (APPENDIX C) shows the Gararu River, formerly the Rio de Pedras, immediately upstream from Curral de Pedras, in the same way as the Gararu/SE town and its river are today.

The last-mentioned map shows part of the two tributaries of the São Francisco, with at least their perennial dimensions due to the dense volume of the main river in the catchment area. On the other hand, the map of the state of Sergipe and part of the state of Bahia (APPENDIX B) outlines the two current Sergipe rivers in question, the Gararu and the Capivara, which contradicts the contemporary cartography of Sergipe. It is important to clarify that this cartographic representation shows the point where the present-day Capivara rises. Despite this, the map does not show this river as its actual course (the former Ilha do Ouro River), but as Porto da Folha. Even so, in order for the map (APPENDIX B) to accurately represent the Ilha do Ouro River (present-day Capivara), it would have had to show its confluence with the tributaries that come from its source, the present-day Rios do Sal and Rios do Cachorro (see figure 5).

The river (Porto da Folha) shown on the map (APPENDIX B), supposedly the Gararu, due to the position in which it and the town are represented on the bank of the São Francisco, is also not the old Rio de Pedras. For this reason, according to current cartography, the source of the Gararu River cannot be in Serra Negra because its tributaries are limited to the municipality of Gracho Cardoso/SE. This is in the northern part of the state, a long way from Serra Negra (see figure 5).

Thus, as the chart of the Comarca de Sergipe Del Rei [1807 or 1809] (APPENDIX C) represents its attributes, it is possible to observe little divergence from the present cartography of Sergipe because, at least, the main points at the mouths of these rivers have been preserved.

As a result, the first navigations across the famous Rio-dos-Currais, from the mouths of its tributaries, were naturally able to enter and explore the surrounding land.

We realise that if the intention was to choose the names of the rivers because of the interests of the time, there may have been a mistake on the part of those who drew up the map of the state of Sergipe and part of the state of Bahia (ANNEX B). In a geographical sense, the river that may have its source in the Serra Negra is the Capivara, like the Sergipe River, although some people call it the Rio do Sal at its source.

At the end of this discussion, it should be pointed out that the information represented on the early times chart (APPENDIX A) is consistent with the contemporary mouths of the two tributaries of the São Francisco, the Capivara and the Gararu. What is striking is the fact that this chart has no date. In view of this, the misconception mentioned above may be confirmed, since it may be the oldest cartographic reference in this case.

With regard to the origin of the name Porto da Folha, after surveying old bibliography from Sergipe, it was not possible to pinpoint it. However, the present work reveals something very relevant in terms of the place's toponymy in relation to its present name. It can be seen that Porto da Folha/SE preserves the name of the old region in which the village of Buraco was located, and consequently, Curral do Buraco was located in that region (Porto da Folha). The relevance of this observation is justified because at no time during the study did the interviewees mention that the town's name had such a remote origin. What's more, we can't rule out the possibility that the expression Porto da Folha was referring, from the beginning, to a river port and gold panning, given that the large region so named was explored by European civilisation.

According to Santos and Andrade (1992), although the town was emancipated in 1896, its name did not come into existence at that time, as its origins are remote. As a result, by analysing the old graphic representation, such as the chart of the early days (APPENDIX A), it can be seen that the name Porto da Folha has existed since the beginning of the territorial expansion of its region, so it has been preserved.

Thus, according to the old cartography of Sergipe Del Rei and contemporary regional geography, it can be seen that, immediately after the discovery of the São Francisco River, its banks began to be explored, mainly through tributary rivers, through which the interior lands of this region began to be explored. It is possible to observe that the circumscription of the present-day Ilha do Ouro village was the starting point for the settlement of Curral do Buraco, which gave rise to the seat of the municipality in question.

It can be seen that the name Curral do Buraco is related to the topography of the rugged terrain, as the area in which it was located is surrounded by steep mountains and hills with undulations ranging from not very intense to difficult to access.

Analysing the chart from the early days (APPENDIX A), it is possible to confirm that at that time (the beginning of territorial expansion) Porto da Folha was already represented cartographically as an area with a large territorial extension . This extension covered several identified points, all bordering the main river in the region's hydrographic basin. Thus, it is possible to see that this name stands out in relation to the settlement once called Curral do Buraco, in terms of merit. However, this does not mean that the current municipal centre did not originate in Curral do Buraco. If this were said, it would be a contradiction, because according to Ferreira (1959), it was Tomaz Bermudes who colonised the land there, founding the corral that gave rise to the Porto da Folha/SE municipality.

However, what is at issue here is the nominal merit, because at no time was it possible to see the famous Curral do Buraco on the map, while the Curral de Pedras was usually represented.

Thus, according to Prado (1919, p. 253-254, emphasis added):

> That was in 1808. Now, it is well known, everyone knows, it is notorious that the family estate, called Porto da Folha, comprised the lands of the municipality that today bears the same name and also parts of Curral dos Bois, Massacará and Geremoabo, ending far above the Paulo Afonso jump, on the bank of the S. Francisco.

For Prado (1919), Porto da Folha was the old Morgado Gomes Castello Branco. It has been noted that a large part of the local population has doubts about the origin of the name, so it is worth emphasising that over several decades Porto da Folha has undergone several changes in terms of the location of its headquarters. In this sense, Ferreira (1959, p. 413) noted that "the main reason for this series of changes was the inconvenience of the town's location, eight kilometres from the nearest port, situated in the village of Ilha do Ouro [...]". The name of the municipality may therefore refer to this harbour, since it is possible that everything started from there because the main access route at the time was the mouth of the Capivara River (formerly the Porteira River), where the Ilha do Ouro village is today. This was one of the hypotheses that the study came closest to regarding the origin of the name Porto da Folha.

According to Ferreira (1959), the famous Curral was founded after the settlement of Buraco. The municipality, therefore, has its origins in the Curral do

Buraco farm. In this way, it can be seen that the settlement referred to by the author may have arisen as a result of the beginning of the exploration of the São Francisco, soon after the discovery of its mouth in 1501. In this respect, in 1522, downstream, the town of Penedo/AL already existed on the left bank of the river.

Thus, even before 1590, the year Sergipe Del Rei was founded, the banks of the São Francisco were already being explored and, after the discovery of its mouth, territorial occupation began with the arrival of Portuguese ships. It is therefore estimated that the settlement of Buraco took place soon after the mouth of the Capivara River was discovered. As for the date, it was not possible to pinpoint it, but it is believed to have occurred even before 1590, when European navigators began to visit the São Francisco River.

NATURE DURING THE RICE CULTIVATION CYCLE

The rice-growing cycle that took place there depended on a natural force. This came from the active whole, a phenomenon present in the bed of the São Francisco River, plus the action of the community, which depended on its survival habits. This whole is embedded in nature.

In this sense, the understanding of nature will be addressed here, so that it will be clarified with the description of the place.

The concept of nature is very complex to express. Philosophers have long discussed it. In the contemporary world, it has been rediscussed precisely because of the environmental issue that has been featured in the written and spoken press, as well as in literature. This work will not delve into this complexity, but what is discussed about nature will no longer be relevant to understanding it. Thus, we hope to achieve the reason for nature, at least in the sense that all living beings and all things in the world can be understood as natural, as long as they are not artificially altered in the fullness of each original state. In view of this, all beings should be understood together as the nature of the world. In this way, it can be seen that "[...] in the 'intimacy' of nature, what prevails is dependence and interaction". (CARVALHO, 2003, p. 74), but this should not annihilate the force that makes them happen, since if this is confirmed, nature may cease to exist.

This is followed by a description of the place, taking into account various natural aspects within the domain of nature, such as the exceptional climate and hydrology of the region, according to Ab'Sáber (2003). This description will focus on the cyclical period of rice cultivation.

In the municipality of Porto da Folha/SE, its entire territory, from north to north-east, has always been bathed by the River São Francisco, facilitating the use of the floodplain for rice cultivation. Until the middle and end of the 20th century, the municipal seat was characterised as the bank of the River São Francisco because of the natural flooding over the large expanse of river plain that stretches from the edge of that river until it reaches the site via the Riacho Capivara, where it forms a single river.

The municipality's administrative area lies in the transition zone between the agreste and the sertão. It is very likely that this area is geologically part of the Primitive Era, as it is possible to see quartzite and limestone belonging to this era. In this sense, Bezerra (1938) noted that the terrains of the Proterozoic period have igneous as well as metamorphic rocks.

As this area precedes Propriá/SE, upstream, it also forms part of the Pedra-oceano geological formation, which according to Valverde (1944, p. 184), is a "[...] stretch of Archean formations up to Propriá, in which the landscape is more rugged than throughout the rest of the lower São Francisco". This is due to the existence of magmatic rocks such as granites on the banks of the River São Francisco, located in the municipality of Porto-Folhense and belonging to the Primitive Era.

The territory of Porto-Folhense fits these characteristics, although the little variability in its relief does not prevent the existence of floodplains such as Ilha do Ouro (Gold Island) embedded in the valley at the mouth of the Capivara. As Valverde (1944, p. 204, emphasis added) rightly observed:

> The region's low relief, which varies from hills to plains, allows the waters to flow gently. From time to time, small alluvial plains form between the hills and the riverbed, which widen as you descend. On the flat part, there are lagoons that drain into the São Francisco. It was in these stretches that the cultivation of cereals, especially rice, took off.

To the right of the mouth of the Capivara is the Ilha do Ouro settlement, on a hill, once a farm with the same name, originating from the site founded there by Gerônimo Costa Taborda, where he established himself with crops and cattle breeding, according to Ferreira (1959).

According to the Topographic Map of Pão de Açúcar/AL (1973) (ANNEX D), the point where the current settlement is located is below the 40 m (forty metre) contour line. Analysing the map, it was possible to see that the hill on which the town stands is surrounded by lowlands from the position of the floodplain of the same name, until it enters a grotto to the right of the Capivara.

At the only point where this gully suffers a brief and not sharp interruption, it occurs at a level far, far below where the houses are clustered, so much so that from the top of the hill to this point the height drops by 18.29 metres.

Having said that, it's important to point out that this slight intersection of the grotto is about 19 metres above sea level. This means that the intersection pointed out here was not high enough to interrupt the occurrence of the factor that determined the island circumstance, although there are those who believe that this intersection was higher in the past than it is today. However, no-one has proved the opposite of what has been said here.

In view of this, it can be seen that in the early studies of the São Francisco River, its torrential flooding "[...] through natural gullies between limestone hills, penetrates far into the land, shredding it into numerous islands". (MESQUITA, [192-?], p. 316). Thus, the hill called Ilha do Ouro was in fact an island, due to the large volume of the river's flow, which reached 13,000 m^3/s, according to Andrade (2004). In this respect, it is possible to note that when exploration of the São Francisco River began (16th century), European navigators discovered the mouth of the Capivara River and were dazzled by this forested island.

The river contributed to the physical formation of the place: the flow of the São Francisco was so great that it periodically overflowed the river plains on its banks. This period occurred between the end and beginning of the year, more precisely from November to March. The floods thus inundated the entire extensive river plain, which was the essential factor for the existence of the lush tropical hydrophilous floodplain forest and the water outcrops present in nature. It was also responsible for the rice cultivation that existed at the time.

In the distance, depending on where you were looking, you could see: in the background, the blue domain of the waters of the São Francisco River; the Capivara Creek, not very wide, but vigorous, from the bank of the river seen in the background, until it turned around in the urban area of the municipality; this in the foreground, on the small plateau surrounded by much higher mountains; embedded in the valley between verdant floral mountains, the extensive river plain that dazzled before the banks of the Riacho Capivara, from the edge of the São Francisco, on the Sergipe side, to the point where the Riacho surrounds the municipal centre.

Porto da Folha/SE is part of the São Francisco River Basin because it is the main river in the region and bathes a large part of this territory. The municipal centre is bathed by the Riacho Capivara, yet until the end of the 20th century, due to the high flood volumes of the "Velho Chico", the urban area was characterised by the banks of this river.

The rhythm of the movement of the waters of the main river attracted the attention of those who observed the place during the cyclical period of this natural phenomenon, because on its extensive river plain there was a relationship between society and nature. This relationship occurred from before the floodplain was flooded until afterwards, so that as the water receded, the number of people working in the rice fields increased.

The Riacho Capivara contained an abundance of water and was perennial from its mouth on the São Francisco until it reached the municipal centre. It is possible to see that the fluvial action of this creek is related to the existing intersection of the configuration of the small plateau where the urban area is located between Rua de Cima (popular name) and Lagoa Salgada. However, it cannot be ruled out that the São Francisco also contributed to the episode through the dynamics of its floods.

When it reached the most extensive river plain in Porto-Folhense, the course of the Capivara was meandering and, from the city's circumscription, it remained perennial until it was lost from sight upstream.

The várzea covered an expanse of river plain that at the time had different names in each private land division, from the Ilha do Ouro várzea, Lagoa Comprida, Lagoa de Miguel Pinto to Lagoa Salgada, among others. Despite this, the Ilha do Ouro floodplain stood out due to its large size, production and personal occupation in rice cultivation.

The relationship between society and nature in the floodplain caused astonishment to any observer when they saw the large contingent of farmers carrying out their activities among the verdant rice paddies on the banks of the São Francisco and Capivara rivers.

Later on, when the rice was ripe, with the sun's rays over the area, it was possible to figuratively see an island in the distance. In other words, the resplendent yellow colour of the ripened rice, set amidst the lush green of the tropical hydrophilous forest that surrounded the floodplain, impressed the observer.

As a result, if any observer allowed themselves to be carried away by natural forces (nature), they could imagine that this was an auriferous island. This was possible if the observer stood with their back to the São Francisco River, at a certain height, not too high, but high enough to see the floodplain and part of one side of the Serra da Lagoa Comprida.

It was noted that there are photographs from the time of the rice cycle, in which in the opening of the landscape you can see the striking yellow colour that is reminiscent of the island mentioned above, since in the background you can see the dense tropical hydrophilous floodplain forest surrounding the valley where the curly rice paddy used to be. The lush green biomass of the forest was present and significant, although the colour of the pictures analysed is not so clear due to the poor quality of the photographs because of the resources used at the time for this purpose. This is why this text is no longer illustrated by at least one of these images.

The long-shaped Lagoa Comprida is in the valley of the mountain range of the same name as an arm opposite the continuation of the river plain, upstream of the Capivara.

Lagoa Salgada's main characteristic is brackish water, and the local population has always complained about this situation. It is strategically positioned to receive water from the Capivara, as well as the Morro dos Carneiros creek. This situation is confirmed by the fact that the less this lagoon receives water from the São Francisco River via the Capivara, the more its water is vulnerable to salinity because the Morro dos Carneiros Creek is a tiny tributary of the Capivara. Since its regime is pluvial, taking into account the region's climate, it results in an intermittent stream. This is due to the climate's water difficulties, which means that the lack of water renewal, due to the absence of rain, contributes to the existence of brackish water in the lagoon. The river plains of the Riacho Morro dos Carneiros were stunning with the rice paddies present and the forest surrounding the valley, as long as the floods of the main river reached them, which was often.

The extensive valley is sandwiched between the mountains of the former Fazenda Ilha do Ouro, Lagoa Comprida, Cal and the intermediate hills below Serra dos Homens. The lush plant biomass bordered the entire valley, so that from the foothills to the slopes of these mountains and hills, hydrophilous tropical forest was found. It also supported the existence of water outcrops and various species of animals that inhabited the region.

The vegetation cover, from the foothills to the mountain slopes, was diverse and varied in height from four, fifteen and even thirty metres or more.

At dusk, the local environment was almost dark, very sombre, due to the marked presence of large trees such as aroeira (Artronium urundeuva), cedar (Cedrella sp.), among others, as well as sucupira (Bowdichia virgillioides) and maçaranduba (Manilkara rufula). Due to the intense exposure of the ingazeira (Ingá Edullis) to sunlight, its crowns developed at the highest level of the vegetation. It was therefore characterised as emergent, a tree that grew above the canopy formed by the tops of smaller trees, but higher than the rest.

Thus, the trees with the highest canopies, which were important for the protection of species that depended on shade for their development, formed the canopy that was surpassed only by the emergent trees.

Analysing the map (fig. 7) at the end of the 1970s, it was possible to see traces of the environment described, such as its natural resources, vegetation and hydrological potential.

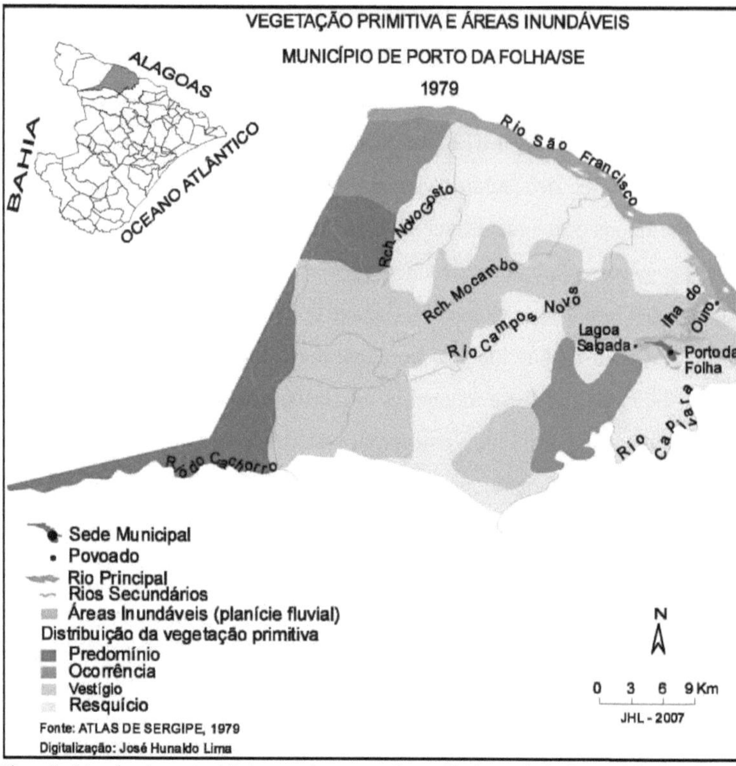

Figure 7 - Map showing natural features such as vegetation and hydrography in the municipality of Porto da Folha/SE.

The araçazeiro (Psidium Spp) was one of the tree species that existed. The shrub layer was characterised by bamburral reaching a metre or more, as well as the presence of caroá (Neoglaziowia

variegata), there were other species of the same genus such as gravatá (Aechmea lingulata). The bitter grass (Andropogon sp.) and the possible existence of macambira (Bromelia lacioniosa) made up the herbaceous layer, although the evergreen grass (Panicum maximum) was observed more frequently on the banks of the river plain.

The flowers of emerging trees such as the ingazeira, among other species, exuded a pleasant aroma that could be sensed in the environment. The reed (Cyperus articulatus), a herbaceous stratum that shared space with the rice fields between the floodplain ponds, exuded a smell that gave the sensation of humidity. The presence of water in the ponds and the strong intensity of the sun contributed to an increase in the relative humidity of the local air, resulting in this sensation.

The marizeira (Geoffroea superba) gave fruit called mari, a regional name, which was eaten cooked by the local farmers. The genipapeiro (Genipa americana), on the other hand, gave fruits rich in iron, which were widely used in the region to make juice. These species, among others, were part of the canopy classification. Also in this classification and characterised as hardwoods, it was common to find the famous sucupira and cedar species, as well as the braúna (Schnopsis brasiliensis). Finally, the trapia tree (Crataeva trapiá) bore yellow fruit called trapiá, which was eaten by birds and farmers alike.

The thorny species were found from the transition zone of the tropical hydrophilous forest to the interior of the caatinga, as it was common to find facheiro (Pilosocereus pyauiensis), xique-xique (Pilocereus gounellei) and mandacaru (Cereus jamacaru).

The marked presence of the natural phenomenon, the flood cycle of the São Francisco River, was a determining factor in characterising the local microclimate, because with the variation in the albedo condition (reflectance of reflected energy in relation to incident energy) in the area of the water mirror and the dense forest, a change in temperature and relative humidity was expected. In this way, the combination of natural phenomena and the presence of these climatic factors formed a microclimate that was different from the semi-arid climate of the caatinga, so much so that, on the way to Lagoa Comprida, there were points of water that emerged under a craibeira tree.

This point is located between the foot of the mountain named after this lagoon and the Capivara Creek. It's important to note that, unfortunately, another environmental impact has been noted at this exact point. This time it wasn't because of the Xingó hydroelectric project, but because of the road linking the municipal centre of Porto-Folhense to Ilha do Ouro. This impact is characterised as irreversible, as the old craibeira tree has been destroyed and the carriageway (Estrada da Lagoa Comprida) where the remains of the outcrops were observed, has been paved over. This impact is not only reflected in the environment, but also in the value of the local culture, since the local people have lost yet another link to the way of life that has been built up over time. Amongst other things, the fact that people who lived through the rice cycle passed by that spot with its emblematic characteristics of the era being discussed here was capable of expressing their feelings. This is because, in addition to the clear water that flowed there, the aforementioned tree attracted people to return home after a day's work in the fields. As a result, the de-characterisation of the exact spot also has a socio-environmental impact. In other words, because the environment has been damaged at that point, the consequence has affected the people, in a way that has deprived them of values (culture, feelings, etc.).

Returning to the characteristics of the environment, it was possible to see that the fauna was distributed according to the factors of the physical environment. For example: in the tops of the emergent trees and the canopy there were hawks (Rupornis magnirostris) and various species of birds with regional names also found among the vegetation, such as briós, parrots, sandpipers, among other birds, as well as the moorhen. In addition to these, two beautiful birds deserve to be highlighted, namely: the magpie (popular name), which is generally black in colour and tends towards green, as well as blue and white, imitates many small birds and used to inhabit mango trees in the region; the sofrê, a bird with similar characteristics and a predominantly yellow colour tending towards orange, but which differs in terms of its size and song.

Among the diversity of animals that were seen, with regional names, the catingueiro deer and the sariema stood out. Other animals found on the slopes included preá, mocó, cágado, etc. Even though these animals were from the caatinga, they were usually found in the tropical hydrophilous floodplain forest because of the transient environment and the abundance of water in the extensive river plains. Thus, many caatinga animals like these entered the forest in search of water. While marmosets (Callithrix jacchus), among other species, were more present in the canopy and usually sought protection from hawks among the branches of the trees, raccoons (Procyon cancrivorus), foxes (Canis vetulus) and armadillos (Euphractus sexcinctus flavimanus) were seen on the ground among the trees.

As for soil classification, in the Porto-Folhense river plain, on the banks of the São Francisco and Capivara rivers, according to the Companhia de

Desenvolvimento dos Vales do São Francisco e do Parnaíba (Codevasf) (2000), Gleissolo Háplico Tb Eutrófico, típico A moderado, textura média, fase tropical perenifólia de várzea, relevo plano. More Gleissolo Háplico Tb Eutrófico, typical A moderate, clay texture, tropical hydrophilous floodplain forest phase, flat relief.

With regard to the local relief, it can be seen that the municipal centre is on a small plateau with an altitude of less than 100 metres, although its level decreases towards the west until it is intersected by the Capivara Creek. Immediately after the intersection, on the other side of the continuation of the plateau is the Lagoa Salgada neighbourhood, whose area is kind of circumscribed by the lagoon of the same name.

Based on this association between the stream or river, the saltwater lagoon, as already mentioned, and the sloping terrain that extends down to the level of the plain, as well as the herbaceous and shrubby plants on the banks of the Capivara, it is worth emphasising the following: it is possible that this ensemble has the meaning of restinga, the name of one of the city's first settlement points. In this regard, it is common in Brazil to hear people talk about restinga in the sense that it is a narrow, elongated stretch of land, vegetated, on the banks of a river, for example. When the Capivara or the São Francisco flooded, this so-called restinga was marooned. This was observed on a small hill in the middle of the river plain, close to the municipal centre.

At that time, when observing the place from a distance, it was possible to see that it had the characteristic of an enclave, since within the semi-arid caatinga there was a dense area of exuberant plant biomass. However, this area was not caatinga, but tropical hydrophilous forest. What Ab'Sáber (2003, p. 145) considers "to designate patches of ecosystems typical of other provinces, but embedded within a domain of a totally different nature [...]" was observed there. In view of this, by letting themselves be carried away by the natural spectacle that existed there, those who appreciated the nature of the place were rewarded with both its fullness and its beauty.

Given the natural setting described above, it is possible to observe the relationship that nature has with each other and note that there has also been an interaction between man and nature, altering the space of the place. On the one hand, man acted on the occupation of space in his own way in the formation of his settlement. On the other hand, nature was in its fullness, vegetating through its environment, taking advantage of the movement of the river waters that were present there and influenced the entire ecosystem (understood as a set that promotes relationships between living beings and environmental factors, such as vegetation, climate, soil, etc.). It can also be seen that the location of a settlement on the banks of rivers is linked to man's need to relate to the environment.

The meaning implied here with regard to man's relationship with the environment can be clarified with the observation by Rolston III (2004, p. 325), "Our well-being, our good living, is a matter of living in sustainable communities, both human and natural".

It must also be taken into account that the river basin is part of nature and must be considered in any study, in its entirety, because according to Rodrigues and Adami (2005, p. 147):

> The river basin is one of the most widely considered spatial references in studies of the physical environment. It currently subsidises much of the legislation and territorial and environmental planning in Brazil and many other countries. However, in most studies there is rarely a precise conceptual definition of this system, which is both hydrological and geomorphological.

Thus, the watershed is not nature, but is part of nature as a whole, and is therefore one of all the things in the world, the whole of nature. For this reason, any undertaking to be carried out after the Environmental Impact Assessment has been carried out and approved, it is important to take into account, as a total space, the surroundings of the watershed, as well as the climate to which it belongs.

CHAPTER 2

THE NATURAL DYNAMICS AND CYCLE OF RICE CULTIVATION ON THE BANKS OF THE SÃO FRANCISCO RIVER IN PORTO DA FOLHA/SE

As we saw in the previous chapter, Porto da Folha/SE was more markedly characterised as the bank of the São Francisco River when it was not suffering from major human interference, as well as with the construction of various hydroelectric dams on its banks. This chapter looks at how rice farming took place in the area when the river was not affected by the aforementioned interventions. It discusses the relationship between society and nature in terms of the natural flood cycle that existed on the São Francisco River for as long as it did. It also deals with the local community's subsistence model, in which farmers kept part of the harvest for family support throughout the year until they started planting again. It also discusses the influence of the rice cycle on local popular culture, which continues to this day.

This cultural relationship is presented as a link between nature and society. This is most noticeable in the way of life of the local people who were more directly involved in rice cultivation. The emphasis is on the sharecroppers, the simplest people who had a unique way of developing.

THE RICE CROP CYCLE AND STRUCTURE

FUNDING

At first, in the Ilha do Ouro floodplain, when it was an enclave of the Sergipe hinterland, as explained above, the land was prepared using the techniques of the local sertanejo. In view of this, according to Cunha (1997, p. 120), "The traditional agricultural techniques with the environment/nature in the Porto da Folha scenario presented as examples, demonstrate a theoretical/practical knowledge of the place that the sertanejo possess".

From analysing the interviews, it can be seen that rice was grown in association with other crops. This cereal, essential to the life of the ribeirinho, was grown on the flooded land, following the decline of the waters, while corn, pumpkin, fava beans, string beans, watermelon, okra, tomatoes and coriander were planted on the combros (small elevations of land). Fish was one of the main additions to meals, since those who couldn't buy it caught it, either in the lagoons, the Capivara River or the São Francisco.

Rice grew regardless of rainfall, but other crops did better when the winter was good, i.e. wetter. Even in years of severe drought, it was possible to see good rice production. The crops associated with the rice fields were also harvested, even though they faced greater difficulties in their development.

The natural phenomenon that occurred in the São Francisco riverbed, the flood cycle, confirmed by some of the interviewees between "November and the end of March[3] ", determined the rice cultivation cycle.

According to Rebouças (1999), soil humidity is essential for the existence of plant biomass cultivated in wetlands. In other words, the fluvial plain of Porto da Folha/SE was inserted in a small strip of wetland, ratifying the author's statement, with conditions for the existence of the rice crop cycle.

This cereal was grown as the waters of the São Francisco River returned to its bed. Rice was sown at the height of the combros, in the lean soil, forming seedbeds that were later transplanted into the lagoons of the extensive river plain when the water level was relatively low.

While the São Francisco River flowed naturally, the river plains of Porto da Folha/SE were flooded cyclically.

When natural floods were unable to flood the entire extensive river plain, water was pumped into the floodplain lagoons using motors attached to rafts across the Capivara River. This was one of the ways that contributed to rice production, guaranteeing the cereal's harvest, albeit smaller in comparison to the natural cultivation resulting from the floods and ebb tides.

The local rice crop depended on the floods of the São Francisco and if the Capivara River flooded it would cause major damage to the crops because its waters were brackish, drastically reducing production.

It was possible to observe that the lagoons on the extensive river plain were

[3] The source has not been revealed for reasons of preservation of rights.

dish-shaped, similar to others on the banks of the Lower São Francisco. This was a fundamental factor for the natural deposit of humus in the lagoons, which always occurred. It is important to emphasise that this humus constituted natural fertiliser, increasing the harvest of rice and all the legumes planted in association, such as maize, beans, squash, etc. in the combros. As Monteiro (1962, p. 27) rightly observed, "The alluvial nature of these soils and their annual renewal mean that salinisation phenomena do not occur within the floodplains".

So, as rice was grown in Porto da Folha/SE and because of the action of nature, there was no need for the sertanejo to use advanced techniques to improve their crops. Not least because the people of Porto da Folha didn't have the technology to use pesticides, nor did they have the purchasing power to do so. Their development model was not based on advanced techniques, because according to

Cunha (1997, p. 69):

> The way of life of the sertanejo in the municipality of Porto da Folha/SE is understood mainly through the values related to the process of using agricultural techniques for survival. [...] From there, traditional agricultural techniques become the instruments to be analysed within the social and environmental dimensions and highlight their contributions to the possibility of a sustainable economic development style for the semi-arid region.

This model was developed mainly by farmers, most of whom were land tenants and sharecroppers, as well as small landowners who carried out the same activities because they couldn't afford to pay a labourer.

As Cunha (1997, p. 108) rightly observed, the local land structure was summarised as follows:

> [...] showing the number of establishments according to total area groups, it can be seen that, classifying establishments as small (less than 1 ha to 100 ha), medium-sized (100 ha to 500 ha) and large (500 ha to 10,000 ha), there is a predominance of small establishments, which make up a total of 3,191 (95.08% of the overall total) and occupy an area of 45,543 ha. Medium-sized establishments account for a total of 150 (4.47% of the overall total), occupying an area of 27,757 ha (30.50% of the total area), while large establishments comprise 15 units (0.45% of the total) and occupy an area of 18,356 (20.17% of the total area).

It can be seen that the average size of the three groups of establishments is around 15 ha for small establishments, 18 ha for medium establishments and 1,223.73 ha for large establishments.

Considering that the former Ilha do Ouro farm had an area of just over 1,320 ha, this confirms that the owner of this property was classified as a large landowner. Most of the other landowners on the extensive river plain of Porto da Folha/SE, in the vicinity of the Ilha do Ouro floodplain, were also medium to large landowners.

The medium or large landowners who had their land in the best natural conditions, such as vast areas of water, made up a privileged social class, and were even wealthy, since just under half of those in the highest category lived in the countryside and were considered wealthy peasants.

It was noted that on only one of these large rural properties owned by owners with different conditions was the plough used by mechanical traction to prepare the land, even if it wasn't extremely necessary. However, this didn't happen every three or four years. It was possible to observe that this activity was only carried out so that the land wouldn't become so compact during rice cultivation.

On the other hand, small landowners, who didn't have much purchasing power, and the majority of tenant farmers and sharecroppers didn't grow grain in this way.

It was observed that they prepared the land before the flood using their own techniques, as Cunha (1997) pointed out. In this respect, "it was natural, the work of nature, the soil was fertilised only with the waters of the river[4] ". Therefore, there was no need to trace the land to decompress it, since it suffered from flooding, so the humus and river sediments helped to mould the soil, contributing to the development of the rice crop. Not to mention the fact that traced land could cause farmers to get bogged down when they were carrying out their activities.

THE RELATIONSHIP BETWEEN THE PORTO-FOLHENSE RIVERBANK AND NATURE AND RICE CULTIVATION: THE INTEGRATION OF ACTIVITIES AND FUNCTIONS

At the outset, it is important to emphasise that, according to Sergipe historiography, rice cultivation originated in Sergipe when it was first planted in 1614 on the banks of the Poxim River. Sergipe (1972, p. 68, emphasis added):

[4] The source has not been revealed for reasons of preservation of rights.

> Rice cultivation was introduced to Brazil by the Portuguese at the end of the 17th century, when they imported the Orysa sativa species from India. However, the historiography of Sergipe states that rice was planted for the first time in 1614, on the banks of the Poxim river.

It was observed that when the natural flooding of the São Francisco River occurred cyclically, which is the main cause of rice cultivation, the good winters helped to increase the local harvest.

Every year, natural floods completely inundate the extensive river plain of Porto da Folha/SE, from November until March. The existence of this phenomenon was so important that, if it didn't occur, the year's harvest would be drastically reduced, even if there was a good winter.

In the off-season, before the floods reached the river plains, farmers prepared the land that could be affected by the natural phenomenon. Tenants and sharecroppers would structure the ditches and their water gates (small sluices) in such a way as to sustain the necessary amount of water for the crops to grow. While sharecroppers were subject to the conditions of rural landowners who were not wealthy, landowners and labourers (many of whom were renters) carried out the activity together, most of the time, using their manual strength and, at most, animal strength. They could also count on the force of nature not to destroy the protective walls of the ditches. This was possible as long as the floods of great intensity (great phenomenal floods) occurred. These floods occurred every ten years or so and when they did, they didn't cause as much damage as the floods usually do, because the river dweller lived in accordance with nature, not the other way round. In other words, at that time, nature wasn't as adapted to man as it is today, just as the São Francisco floods today occur in its lower reaches, controlled by the various dams, and finally the Xingó dam.

In order to produce on their rural properties, wealthy landowners relied on the manual labour of sharecroppers and the paid labour of rural workers who worked for hire. Despite this, one informant said that he used to use a tractor on Várzea Ilha do Ouro to lay out the land, but it was clear that "So I only removed it... because the land was very compacted. So he only removed it every three years, every four years, that was Dad's expense[5] ".

[5] The source has not been revealed for reasons of preservation of rights.

The Várzea do Araticum was also important for the town, as the farmers, sharecroppers, had their activities divided between this Várzea, Ilha do Ouro and all the lagoons on the banks of the Capivara, adjacent to the municipal centre. This Várzea is to the north-west of the village of Ilha do Ouro, on the banks of the São Francisco, so it is further away from the town than the Ilha do Ouro Várzea.

Even so, the extensive river plain adjacent to the municipal centre wasn't so vulnerable to flood damage, as it is set in a valley, protected from the river's strong current. In short, even the great floods didn't prevent good harvests. There were no houses in the area of the river plain, just the structure of the lagoons, and as soon as the waters advanced along the bed of the Capivara, they found the water gates open. As a result, the lagoons with adequate infrastructure were filled in.

Activities such as clearing the ponds with a scythe, hoe, mattock, pickaxe, etc., to carry out the swidden, were carried out by the men. They were also the ones who carried out so many other activities, the heavier or more difficult ones such as tapageiro work, cutting, churning and transporting the rice. In the region, the tapageiro was the man who controlled the water in the lagoons by handling the mobile instruments of the water gates, boards hoisted up by two concrete bases for this purpose. These people were equipped with nets to catch fish from the lagoons, as this was the crucial point because there was only one exit to the river.

As the water only passed through the water gates, the ditch at this point was usually a little deeper because of the current of the water, both at the entrance to the floodplain lagoons and at the exit back to the river. There were therefore around four tapage workers who also used bamboo mats to prevent fish from passing from the lagoons to the river, so they dived down several times a day to install the artefacts and, above all, to plug the holes that appeared in the mats. The techniques they used were rudimentary.

This fishery, which according to Monteiro (1962) was organised by landowners who kept the floodplain cauldrons as fishponds, took place before the rice was cultivated.

With regard to rice cultivation, it was noted that the water accumulated in the lagoons when the river flooded, from "April onwards, it would start leaking to make the beds, and then the planting[6] ".

Thus, as the flood waters receded, the tapageiros controlled their return to the river, giving the farmers the chance to sow rice in the lagoons.

With regard to sowing, at first the local farmers prepared the rice in order to select the best seed for planting and left it to soak overnight. The next day, they would scatter the seeds on the mud of the ponds, at the height of the hillocks, in the form of a seedbed.

One of the old varieties of rice grown in the Lower São Francisco was the chatinho species, whose grain was rounded. After this species, over time, purple cane rice was planted, which had a crooked shape and needle-like grain, among other varieties.

After about thirty days, the rice was ready for transplanting, which took place as the waters receded, as the volume of the river dropped due to the floods. Then the planting began. This activity occupied the women, but it wasn't difficult to find their children and blood relatives helping out. While some say it was inhumane for children to take part in this activity, it has to be said that the freedom the children found in that space close to nature allowed them to help their mothers by playing with each other. The water, the diversity of birds and some of the local native fruits such as pau de lagoa (a regional name) were all part of the children's games.

There were usually teenagers (around 14 years old) pulling up the rice seedlings in the beds. Meanwhile, their mothers and other women carried the bundles of seedlings to the swamp to transplant them. The younger children played in the shade of the trees with corn dolls and cowries, among other natural things. It's important to emphasise that the presence of children there was a way for mothers to educate and take better care of their children. In the end, they didn't stop playing and were happy to have their mothers around, and they were always attentive to their children to avoid any disturbances. Normally, when the children left school, they had lunch and went to the ponds where their mothers were, while the younger ones stayed at home with their older siblings who looked after them.

Women's and children's activities were usually household chores, and they also worked as craftspeople embroidering fabrics, making fishing nets, among

[6] The source has not been revealed for reasons of preservation of rights.

other tasks, as well as pounding rice in the pestle to separate the husk from the grain. As for the children who lived in the countryside, it was common for them to fetch water by donkey from the river, from ponds or lagoons for their own consumption and, usually, for animals. The donkey was harnessed with a cantle and hoisted on two wooden crates with their outer sides open, except for the front and back, and the child usually rode along.

Once the entire space of the extensive river plain was occupied with rice planting, it took around three months for the harvest to begin.

Meanwhile, in the floodplains, it was common for farmers to live there temporarily, in small huts made of rustic wood, from the roof to the walls, which were made from rice straw or reeds. It was an easier way, in a way, to protect the rice paddy from the birds that flew around the plantation from early morning. To do this, many farmers would set up everything from large cloth dolls to upside-down cans with jingle bells tied to long cords hanging from stakes, and when they didn't work out, they would set off firecrackers to scare the birds away. All of this had one purpose: to ensure the development of the grain, especially when it was ripening. If this wasn't done, the farmers feared that these birds would cause damage to their crops, as several flights of birds would fly over the entire rice field.

The brió, among other species, was common in the area. When the flock landed on the curly rice, all the people present had to do was pull the cords or ropes attached to the devices and the birds would be startled, so that one part would go towards the forest and the other would insist, always requiring the presence of people for this purpose.

Table 1 - Porto da Folha/SE: fruit and vegetable production in tonnes (T) and thousand fruits - 1980

Product	Quantity in (T)	Quantity in thousand fruits
Pumpkin		6
Watermelon		2
Tomatoes	1	

Source: Ibge. IX General Census - 1980

By force of nature, among other natural factors, such as the intense fertility of the soil due to the humus deposited by the river in the floodplains, the rice grew quickly. At the same time, it was possible to continue planting where the waters receded last and to harvest from the first spots planted at the beginning.

In Porto da Folha/SE, it turned out that rice growing was not just a monoculture, but was associated with other vegetables, beans, maize, broad beans, etc. Rice cultivation in the area predominated in the shallows. The entire area of the combros was taken up by these vegetables along with horticultural crops such as okra, coriander, potatoes, watermelon, tomatoes and various species of pumpkin. Table 01 shows the production of some of these products.

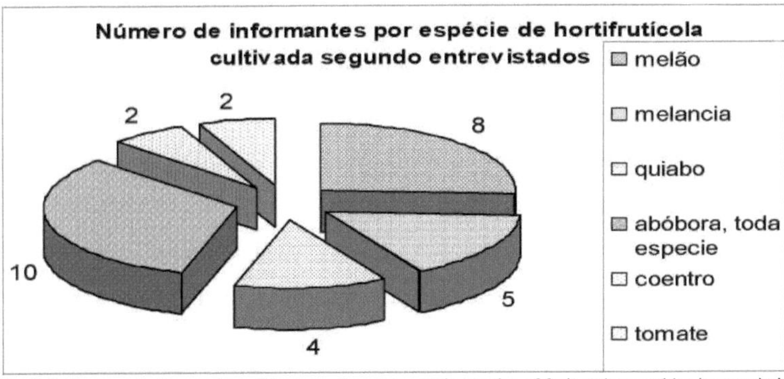

Graph 1. Represents the number of farmers who grew certain species of fruit and vegetables in association with rice cultivation .[7]

The Brazilian Institute of Geography and Statistics (IBGE) took into account the quantity produced, represented in table 01, in tonnes and in thousand fruits, so quantities below this level are not presented in this case. As can be seen in graph 01, several other types of vegetables were grown in the area.

The rice harvest began in August. It was an arduous manual labour job to cut and gather the grain, so it was the men who carried out this activity. It was observed that some farmers cut the rice more sparingly because planting took place later due to the area to be planted taking longer to reach a level suitable for cultivation. While others cut faster because they had planted more and earlier, due to the possibility that the water level gave them. Thus, rice cutting ranged from moderate to fast. It was taken both in bundles and in sacks to the yards set aside in the middle of the plantation, in dry spots well away from the wet area.

The rice rushes spent the night sheltered from the sun under wooden and straw structures. These structures differed from the huts in which the farmers

[7] The source has not been revealed for reasons of preservation of rights.

camped to tend their crops because their openings were open. Only the upper part was covered with rice straw or reeds from the lagoons on rustic wooden bases without walls.

As soon as the rice rushes were gathered, the farmers began beating, the process of separating the husked grain from the straw. The place where the pounding was done varied, because in some cases if the amount gathered under the haystack wasn't so intense that it could be done right there, they preferred to do the pounding there because of the intense sunny days. In this way, it depended on the interest of each mutirão and the condition of the service.

During part of the development of the rice crop, the maris trees had ripened their fruit. This natural phenomenon collaborated with a large number of farmers who picked the maris from under the trees to eat after the fruit had been cooked. According to the Northeast Development Agency (Sudene/Adene) (2005), 1980 to 1984 were years of drought. In such drastic situations, children were often seen collecting these fruits to take home. After being boiled in the evening, they were eaten the next morning after being broken with a stone or other object. The seed is similar to an almond, but smaller and harder. The fruit was fleshy and many people ate the outer part, depending on their food needs. The flooding of the São Francisco river was another natural phenomenon that contributed to the seasonality of the fruit flora.

From sowing in the beds, to transplanting in the ponds, to cutting the rice, they heard everything from farmers' songs through the rice field to regional modinhas. When shaking the rice straw, whistles were also often heard. They said that this was a custom intended to attract the wind, when it was not very intense, because the intensity of the wind helped them to separate the rice more quickly. To do this, the person positioned themselves against the direction of the wind with a can at head height to dump the rice in the yard, so that the husked grains fell next to their feet, while the straw, being lighter, fell in another place downwind. Even though this is a primitive process, this is how the straw was separated from the rice that was still in the husk.

Straw has always been used for animal feed, so it was stored in the hope of protecting cattle from long droughts, regardless of whether it was summer or winter, and those who didn't have this type of feed often had their herds reduced.

The grains, still in their husks, were bagged to begin measuring the harvest. The measurements used in the region were commonly salamim and alqueire (rice measurements in litres) which, in the region, mean 10 litres (or 8 kg) and 240 kg respectively.

When the paddy rice was measured, the farmer, in this case the sharecropper, the landowner or the manager of his property were present. Then the harvest was divided up, with half going to the sharecropper and the other part to the landowner. Maize and beans, on the other hand, were not divided into halves, but into quarters, with three parts of the harvest going to the farmer and one to the landowner.

Other products associated with rice, such as the fruit and vegetable harvest, remained for the farmers, although some thanked the owner of the land by offering him a few units. In this sense, it was possible to observe in the case of Fazenda Ilha do Ouro "the following, corn... rice and corn had this law of... dividing. Now, a watermelon, a pumpkin, [...] these things, people would come to the house and give them as presents, Mr Antônio, I brought you a watermelon, you know? He'd let people plant it[8] ".

Fishing on rural properties was dependent on two factors: if the farmers fished for their livelihood, the owners gave them permission. However, if they opted for this activity in order to commercialise, the owners of the ponds demanded money for it. However, if the fishing was not carried out on private property, but in the rivers, people were not prevented from fishing.

The harvest was transported by ox carts, unlike the way the farmer drove the cut rice to the yard. It was also transported on the backs of animals such as donkeys or even by canoes across the Capivara towards the São Francisco River, when the rice had already been bagged, making it easier to transport it to the warehouses at Fazenda Ilha do Ouro.

It was also common for people from the municipal centre to travel by canoe to the city of Própria/SE, the main commercial centre in the region at the time, to take part in the fairs.

Ox carts were usually pulled by a pair or more of tired oxen and driven by

[8] The source has not been revealed for reasons of preservation of rights.

carreiros, people who drove the animals ahead or on top of the load, equipped with long, often pointed sticks or an iron device at one end.

After bagging the paddy rice, there was a lot of seed left over in the yards. This attracted flocks of birds that fed well, sharing the space with the chickens raised by the farmers and with waterfowl (a wild species that was very common in the region). From then on, the birds were no longer frightened off because they no longer posed a risk to the crop. In fact, this was often the case if the crop had already been fully harvested there. These birds are the ones that used to fly over the floodplain all the time when the crops were ripening.

The sharecroppers dried their share of the rice on the pavements of their homes and in the neighbourhood. During sunny days, the husked seeds were exposed to the sun, so they dried very well and quickly. The drying of the landowners' produce was no different, although they had better structures such as space to spread the rice on the ground, and they also had large warehouses to store the grain.

Once dried, the paddy had to be processed before it could be consumed. To do this, farmers who were not willing to give another part of their produce to the owner of the rice mill, in addition to the half that they had already given to the landowner, would first pound the paddy rice in a mortar and pestle. In this way, a large part of the riverside community benefited from their produce. However, this didn't happen so easily or quickly, it took time. As necessary, each person in their home, either with the help of family members or relatives and friends, left the rice husked and ready for consumption.

On the other hand, processing by rural landowners took place in old and obsolete pestle factories. However, when these still didn't exist, the grain was also husked by hand by the estate's employees.

The figure (no. 8) below shows one of the first rice-processing machines in the region. Even today, the obsolete machine can be found in the ruined building where the old rice mill used to operate on what was then Fazenda Ilha do Ouro (now the town).

Figure 8. Machinery from the old rice mill on Ilha do Ouro. Source: photograph by the author, January 2007.

On the left, the picture (no. 8) depicts the old machine that worked with a hand crank. On the right, you can see the compartment of the machine that distributed the milled rice and sorted it.

in sequence, from normal rice to some types of cereal sherry such as coarse, medium and fine.

While most of the rice produced by the sharecroppers was used to support their families throughout the year until the next harvest, the landowner's share was stored and sold when the price increased. Trading took place both in the municipal centre and in the city of Propriá/SE and, from there, the produce was exported to the Northeast or other Brazilian regions.

This corroborates what Silva (1978) rightly observed: the rice found on the tables of the poorest was also part of the diet of the majority of wealthier consumers.

According to Ab'Sáber (2006), the way the riverside community related to nature, even though it was between urban and rural areas, can be compared to the way of life of human groups that were not so aggressive towards the environment and used techniques adapted to the rhythm and way of being of nature. In other words, the techniques used in this model did not have the potential for environmental degradation that contemporary technology has, but productivity was not low. What's more, most, if not all, of these techniques were adapted to nature, since the water was deposited in the lagoons and the intense fertility of the soil in the area's extensive river plains was a result of nature.

The analysis of table 2 shows that there was no significant use of advanced techniques when it came to agricultural work and the cultivation of productive land. Thus, according to the author's conception and the context juxtaposed here, the above assertions are ratified, since the table shows that during the rice-growing cycle, the vast majority of local establishments used human labour to carry out agricultural work. This was followed by animal power. On the other hand, it was observed that the community did not use mechanical power, or the combination of animal and mechanical power. Despite this, this study found that only one establishment, mentioned above, used mechanical power, but it seems that this power was not entirely necessary.

Table 2 - Porto da Folha/SE: establishments according to the labour force used - Agricultural work

Municipality -	Human -	Animal -	Mechanics -	Animal and mechanical
P. da Folha	1369	10	0	0

Source: Ibge. 1960 Agricultural Census

If you look at figure 9 below, you can see how important rice cultivation was for the community. The picture also succinctly portrays the way the local people developed during the rice-growing cycle.

Given the expression of the religious value of the local community, according to the picture (fig. 9), the main church represents its economic, social and cultural development.

In the engraving (fig. 9), on the left-hand side of the centre of attention, which is the photograph of the image of Jesus Christ, you can see women transplanting rice into a pond. Immediately afterwards, also on the left, is the figure of one of the greatest expressions of the backlands, a donkey with a possible load of water for human consumption. Above the representation of rice cultivation, we see school education, a crucial factor for socio-environmental and economic development. Above, you can see the fish that used to be in the São Francisco and Capivara Rivers, as well as the lagoons on the extensive river plain. Still in the centre, a little higher up, on one side you can see the figure of a cowboy herding cattle and on the other, men preparing the land for crops such as rice.

Figure 9. Representation of the development of the Porto da Folha community printed on the altar wall of the Nossa Senhora da Conceição Church, Porto da Folha/SE. Source: photo by the author, January 2007.

On the right-hand side of the picture (no. 9), you can see social equality, since the printed features show people of different descents, such as the white man from Europe, the black man from Africa and the native Indian from Brazil, certainly from the Xocó tribe, but all carrying out their activities together without contempt. Immediately to the right is the female figure treading the rice in the pestle. Just above is a representation of the main means of transport at the time, the ox cart, and the figure of the carreiros.

In the centre background, lastly, you can see the Ilha do Ouro region and a possible craibeira tree from the tropical hydrophilous forest that surrounded the entire river plain of the place, especially the Várzea da Ilha do Ouro.

About the aforementioned tribe of Indians, according to Souza (1944, p. 43), from the mouth of the São Francisco:

> At a distance of thirty leagues upriver is the Mission of St Peter, where there is an Italian Capuchin who leads three hundred very indolent Indians, who live off fishing and very little manioc, which they plant on this small island of St Peter, which is half a league long and four hundred fathoms wide.

Today, the tribe of Xocós Indians from which the aforementioned mission originated can be found on the island mentioned by the author. It is therefore possible to say that this tribe is the last indigenous remnant of Sergipe. According to Mott (1986), it had a population of 300 Indians in 1808. Today, the tribe has been on the old island of São Pedro since its earliest days. Due to human activity on the bed of the River São Francisco, this island has now been de-characterised in the same way as Ilha do Ouro. However, when the Xingó Dam releases a large volume of water, the territory of São Pedro Island is still marooned.

LABOUR RELATIONS IN THE FLOODPLAINS AND THE SUBSISTENCE AGRICULTURAL MODEL

Because landowners had a lot of land, they prioritised cultivating the most important areas, such as the river plains. Since rice was grown there on a cyclical basis, this made it unattractive for landowners to keep residents on their properties without work for the rest of the year, since in some cases the work was done by salaried workers. This category could better be called salaried due to the labour system because, most of the time, people were not hired, nor did they receive fixed amounts as real salaries. In addition to the existence of farmers in this situation, others didn't even receive money for their work.

It is important to distinguish between the two terms above: salary can be understood as the fulfilment of a duty and payment in monetary value from the employer to the employee. Remuneration, on the other hand, encompasses both the meaning of salary and can go beyond it, such as receiving bonuses for services rendered and the exchange of per diems between workers. It was observed that when some workers carried out certain activities on the farm, they received cotton or another type of vegetable in return.

Thus, the amount paid to a person for a particular service was variable, as a tapageiro did not receive the same pay as a hired labourer pulling up a stump with a chibanca (an iron instrument identical to a pickaxe, but smaller, widely used in the agricultural activities of the sertanejo). Even farm managers weren't paid the same as any other labourer. Most of the time, there was paid labour because a salary implies that it is a regular form of payment that stipulates at least a minimum monetary value paid to workers in general. This was not the case for most of the local farmers.

It was observed that during the rice cycle, the cowboys on the farms in Porto da Folha/SE were generally rewarded with one out of every four calves that were born, while the hired labourers picked corn and cotton to earn their daily bread.

The hired labourers were men or women who sometimes received payment in cash, albeit derisory. They were also rewarded with a certain amount of harvested produce such as corn, beans or even cotton, which was usually insufficient for the labourer to survive. They also simply exchanged their labour for other people's favours in tending to their crops.

Most of the time, land preparation for rice planting was the responsibility of the sharecropper on the piece of land that belonged to him. On very small properties, it was hired labourers who did the work or the landowners themselves.

According to Andrade (2004, p. 142), "Wages vary according to the number of arms available and the type of bush to be cleared. When the clearing is finished, these wage earners are dismissed and start earning their living from other activities - fishing, hunting, farming - which come up sporadically."

The relationship between landowners and sharecroppers was unlike any other, as there was usually no monetary remuneration. The sharecropper's labour was rewarded with half of the produce he produced himself, from sowing to harvesting, as well as, in some cases, being responsible for the initial preparation of the land. The seed, selected for planting, and the land belonged to the landowner.

In the Ilha do Ouro floodplain, the area (plot) of land divided to each sharecropper was economically viable for each farmer and there was a lot of competition to take part in this system. Because of this, when a sharecropper stopped participating in the system, he would appoint someone to take his place. It was usually the son who took over when the father's time came to an end. At Fazenda Ilha do Ouro, one hundred and seven families worked and lived with the owner's consent. There were times when twenty sharecroppers planted in a single lagoon. "They had lagoons [incomprehensible] in one lagoon, they planted twenty, twenty people... twenty families. Now, everyone knew where they belonged and where they didn't belong. There were 107 families who worked all over the floodplain[9] ".

The sharecropping system was not restricted to the rice farmers and the owner of the floodplain, but also took place between the players in this form of agricultural partnership throughout the river plain on the fringes of the municipal centre of Porto da Folha/SE. This system was also observed in the Lower São Francisco.

It should also be noted that many landowners paid the sharecroppers, so that the sharecroppers were paid at the end of the harvest. This happened if the sharecroppers were interested in loans during the rice-growing period, and they could even repay the sharecroppers with part of their inheritance. While the women and their children were busy transplanting the rice, it wasn't hard to see that the men, their husbands, were looking for other activities to support the family while they waited for the time to do the hard labour. It's important to note that not all family masters acted in this way.

The above statements are in line with what Andrade (2004, p. 142, 143) observed:

> As sharecroppers need to buy food and don't earn a salary, it's common practice for landowners to finance them by charging interest of 6 to 10 per cent a month, to be paid at harvest. The sharecropper is responsible for looking after the rice field from transplanting to harvesting, for an average of three months. Once the crop has been harvested, the rice needs to be threshed so that it can go to the mills to be husked.

This different relationship, commonplace in the area, was emphatic due to the way in which the sharecroppers of the floodplains related to the traders in the city, who owned the bodegas of the time. Many of them bought in instalments and paid with their share of the rice after the harvest.

Most of the time, in terms of monetary value, the catch was commercialised by the owner of Várzea Ilha do Ouro, who sold it outside the area, such as to markets in cities in Alagoas and Sergipe. The people who caught the fish for the owner to sell to traders were paid in cash.

Amongst the tapageiros, other people were involved in fishing in the lagoons. While the tapageiros held back the fish with bamboo mats set up at the points of the water gates, the other fishermen fished inside the lagoons dragging extensive

[9] The source has not been revealed in order to preserve rights.

nets.

After this activity, the buyers of the fish paid in kind to the women who processed and salted the catch. They also paid the owner of the ponds for the fish, and the fishermen were paid by the owner or simply received part of the catch for their activity. The fishermen themselves then sold the fish they caught in the town.

As for the agricultural model that was developed and finally developed in the floodplains in relation to rice cultivation, it is important to emphasise that, while on the one hand, family units in comparison to large rural properties, according to Lima et al. (2002, p. 14) "[...] better ensure environmental preservation". On the other hand, depending on the context in which the crop was grown, it's very likely that their model is even more environmentally viable than various irrigation systems, as well as those that rely exclusively on the use of pesticides to produce.

As Costa (2003) observed, family farming produces all year round in a diversified way and is intended to provide for the family, as well as selling some of the surplus. This was not the model that existed in the riverside community of Porto da Folha/SE, as it depended on rice cultivation to survive.

According to Oliveira (2001, p. 51):

> From Marx's statements, we also have to understand that on small peasant farms, part of the agricultural production goes first and foremost to the consumption of the producer, the peasant, as a means of immediate subsistence, and another part, the surplus, is commercialised in the form of merchandise. In this production unit, as in capitalist ones, there is the presence of differential income obtained from more fertile or better situated land.

In Porto da Folha/SE, the surplus production was commercialised by the landowners, unlike the relationship that the sharecroppers had with local commerce. The sharecroppers exchanged part of their production for other types of products that were also substantial to life. For example, they exchanged a certain amount of rice for a certain amount of flour produced from manioc grown in the drylands, and so on.

It would be a luxury for anyone to exchange any percentage of their production for material goods such as an old sewing machine, even though it was necessary for mothers to make simple shorts for their children, which was common in local families. Furthermore, exchanges of this kind hardly ever took place. When it did happen, it was usually done by people who did not, exceptionally, subsist on this type of culture. In other words, they had another livelihood and didn't depend solely on the fields or growing cereals.

According to Oliveira (2001, p. 51):

> [...] absolute rent comes from situations in which surplus value is realised in addition to the production price, i.e. a monopoly price raises the price of the product above the production price. Marx points out that the value of peasant products is, as a rule, above the price of production, due to the predominance of living labour materialised in these products.

This conception shows the reality of the commercial relationship that rural landowners with greater purchasing power had with the place or elsewhere.

As a result, it can be seen that there were two social classes with marked distinctions, namely: sharecroppers and landowners.

Although the survival of sharecroppers varied according to the amount of land cultivated, among other situations, the family contingent promoted this variability in terms of better harvests. As a result, there were sharecroppers with better living conditions than those of the same class. Thus, whoever had the most children among the young and old resulted in a greater labour force, so that, in place, this meant power. That said, as the number of family members in the agricultural occupation grew, the production of the crop increased due to the large labour force employed there.

The two main social classes mentioned above emphatically help to distinguish between two important forms of agriculture developed as a result of the culture of vazante on the banks of the São Francisco River in Porto da Folha/SE. It can be seen that rural landowners produced agricultural surpluses, as they waited for suitable and opportune moments to commercialise their produce. The sharecroppers, on the other hand, worked for the survival of their families, as they depended on subsistence rice farming.

Because of this, it is important to emphasise that according to Lamarche (1998, p. 70, emphasis added) "The project of the **subsistence** unit is, first and foremost, to preserve structures that allow the survival of the domestic group; it is obvious that that of the **peasant** establishment cannot be summed up in this". In the last model, according to Lamarche (1998), those who practise it aim to progress in terms of their assets, because they don't just farm to survive.

Meanwhile, for Cèpéde, Houtart and Gond (1967, p. 352), these three objectives are in the interest of the peasant to increase his production, not to want without power and to change the structure of the economy:

> [...] they are closely linked to each other, and to disassociate them is a theoretical speculation, since in reality all three are at the same time cause and effect of a complex situation in which the economic goes hand in hand with religion and customs, which is proof that culture is always more or less integrated.

For the authors, this triple objective is fundamental to the development of subsistence production.

For Lamarche (1998, p. 76, 77), this means peasant or subsistence farming:

> The establishment's production (most of which cultivates small areas, often less than ten hectares) is destined for family food and only a very small part of the production is commercialised. [...] Likewise, these producers are not very sensitive to the notion of agricultural family reproduction (wanting their children to continue farming, using any savings to improve production structures, etc.). On the other hand, most of them believe they can live for around a year in complete self-sufficiency. All this leads us to believe that these producers operate more on a subsistence and survival model than on a peasant model.

Given this concept, it is possible to see a relevant point when the author states that production is aimed at supporting the family. Despite what has been said above, the development model of the porto-folhense riverside community, while it existed, was closer to subsistence.

It is therefore possible to say that the model of agricultural development pursued by rice farmers in the riverside community of Porto da Folha/SE was family, peasant or surplus farming. However, if each of these models were not restricted to their particular definitions. Given this, the model developed in the area by the sharecroppers, due to its particularity, was observed as a unique agricultural model, despite the fact that there was a family, peasant and surplus farming relationship there. However, this was different from the emphasis given to the meaning of each of these models, as observed by Lamarche (1998).

According to Lamarche (1998, p. 311), "Although the survival function is very present in the peasant model, it is impossible to reduce it to that; it is deeply rooted in a desire to conserve and grow the family heritage".

It's true what the author identified in relation to survival or subsistence, this development model is at the extreme end of necessity, so much so that in the case of the Porto da Folha/SE community, the vast majority of rice farmers didn't own property.

In this case, there were two distinct models of development: one that took place in the mould of the sharecroppers' lives, and the other in the way rural landowners, especially the wealthy ones, marketed their produce.

Taking into account what Lamarche (1998) observed, there are no absolute models, as he made clear the possibility of one model being present in another.

As such, this concept ratifies the model of development that was practised by the riverside community on the extensive Porto-Folhense river plain, on the banks of the São Francisco River, based on rice cultivation.

A HIGHLIGHT OF PORTO-FOLHENSE CULTURE DUE TO RICE FARMING

The traces of popular culture have long endured in a marked and notorious way in the place due to the rice culture that was developed on the extensive river plain.

One of the strong characteristics of the human group present there was the way in which they managed to survive in the face of so many difficulties, as Souza (1944) observed, such as the perishing of cattle over the years, as well as the long-lasting droughts that affected the vegetation. The local process used from cultivation to the processing of paddy rice for food was a custom passed down from generation to generation. The means of survival of the sharecroppers, the rice farmers, had its identity in this custom.

The cyclical and annual habit was present in the way the local people cultivated rice, also depending on the natural cycle of the São Francisco River floods. Values were established at the moment when fathers, mothers and children were present, so that the younger generation used the lessons they had learnt about growing the cereal.

Thus, while fathers taught their children how to prepare the land for planting rice, mothers taught them how to sow the cereal in the beds and transplant it into the ponds. Fathers also taught them how to separate the straw from the rice (pounding rice). The children also learnt how to husk the rice in a mortar and pestle, because the vast majority of local families couldn't afford to pay the mills to carry out this process.

This procedure was usually carried out on a rigid wooden instrument carved in half on the outside, known as a pestle, measuring just over a metre and a half in height, with a diameter of around seventy centimetres at the two ends and a cavity of around two palms. In order to carry out activities on this instrument, it was necessary to use pestle hands, other instruments that were around eighty centimetres long, with a slight cavity in the centre for manual support, and around twelve centimetres in diameter at the carved ends. All these measurements vary according to the height of the person handling the instruments or the utilities they require.

You could often hear the sound of rice being pounded in a mortar and pestle on Restinga, one of the city's first streets, and henceforth on Rua de Cima, also known as Rua da Restinga. This sound was produced in the process of husking the rice. To do this, one end of the pestle rested on the ground. The movement with the pestle hand was done in such a way that, when it was raised to head height, whoever was carrying out the activity would pour their strength, with fists clenched around the instrument, into the amount of husked rice inside the pestle, resulting in the aforementioned sound. It was observed that the vast majority of farmers carried out this process by hand, in a rudimentary way.

In this relationship that the people had in passing on knowledge to their future generations, it is possible to confirm the existence of customs found in the place, such as the consumption of typical foods emphasised at certain times. In this regard, we can highlight the manuê de arroz, a cake whose main ingredient in Porto da Folha/SE was the rice grown in the municipality's floodplains.

Mothers taught their children how to make the sweet delicacy in their own way, so that after the rice had been peeled and the seed husk shaken off, it spent a certain amount of time soaking in water which, after being drained, the cereal was pounded in a mortar and pestle until it became dough. Once this was done, the seasoning was added, the cake was baked and then served to the family. It's important to emphasise that given the difficulties that the local people also faced, as already mentioned, this custom may have had a certain exclusivity, that is, it was usually only carried out during the Christmas and New Year festivities, in an emphatic way. "Yeah. Only at the Christmas party, it [manuê] is so dependent on the year that we only do it at parties[10] ". Although this event was emphasised during festive times, there was nothing to stop some wealthy people from making the cake whenever they wanted to eat it, but we mustn't forget the great distinction between the two social classes characterised above.

With the floods at the end of the year, the festive season, Christmas and New Year, people entertained themselves in the waters of the São Francisco that washed over the municipal centre through the floodplains and the Capivara River.

In the photograph (figure 10) below, a light stands out in the background, where the famous Rua de Cima (Upper Street) is popularly known, and in the frame formed by the local residences we can see the Praça da Matriz (Main Square) populated by the New Year's Eve party crowd. It was at these parties that all the family members, now without distinction of class, race or colour, met and celebrated as a community.

Figure 10. Praça da Igreja Matriz during the Christmas and New Year festivities. Source: author's photograph, January 2007.

The Praça da Igreja Matriz (Our Lady of the Conception), as shown in figure 10, is the site of the festive gathering, which took place a long time ago and continues to this day. Many delicacies were sold on stalls set up in the middle of the crowds and the people's entertainment was marked, among other things, by the disputed rounds of the old Zé Malfeito curri, set up in the square. Nowadays, you can still see the curri, but without its striped cloth cover. As can be seen in the

---

[10] The source has not been revealed for reasons of preservation of rights.

picture (no. 11) below, the curri is on the right-hand side of the square, where pedestrians pass by.

Figure 11. Curri de Zé Malfeito in time for the Christmas and New Year festivities. Source: photograph by the author, January 2007.

In terms of its structure, it was observed that the curri is a device made up of pieces of wood and rebar that are structured on a large support that is ploughed in the shape of a roller, so that it rotates around an axis. Surrounding this wooden support, a structured plank path is hung from the horizontal pieces, supported by the largest piece, as shown in figure 11, on which wooden seats with simple upholstery are fixed. Its conical upper structure was usually covered with fabric.

Another important cultural factor that has resisted the transformations of time and space is the contemporary cowboy festival, since from its beginnings to the present day, the development process of the local people has had strong links to these cherished values in their lives. Although this festival, called vaquejada, seems to have no connection with rice culture, this is not the case because it was possible to observe that the event originated on the old Ilha do Ouro farm.

What links the vaquejada to rice cultivation is the fact that while the Fazenda's floodplain was not being used for its main purpose, the property's cattle were usually grazing there. On the other hand, on the eve of cereal cultivation, the animals were led off the farm by their cowboys. After the harvest was done, they would cowboy the cattle that were in the solta, the caatinga of the high drylands and shallows among the pingurutos. In this way, it was possible to find the cattle again, facing the great obstacles of the closed forest. They also found the wild cattle that were raised in the caatinga because it was so difficult for the cowboys to find all the animals that were scattered there.

It was this rural activity that gave rise to the "pega de boi no mato", which still exists today, even if it has been improved, but its original value remains. The meaning of "pega de boi no mato" is: as the cattle were loose in the bush (caatinga), the cowboys of the Fazenda had to cowboy the animals and tie them up in the corral. Until this fact was confirmed, many of them had to go through great adventures, immersed in the semi-arid vegetation; thus, several times they had to tie up the bull to lead it to the corral; as a result, the bull catching arose in the caatinga, natural vegetation of the place, this was part of the organisation of a local cowboy; from then on, each of the wildest bulls was released from the corral for the cowboys to catch, those who managed to bring the bull back to the corral were rewarded. In this way, the culture of the sertanejo cowboy, from Porto Folhense, was formed, which had its origins in entertainment and, nowadays, is even spoken of as a sport.

It was at Fazenda Ilha do Ouro that the vaquejada originated in Porto da Folha/SE, with the aforementioned bullfighting in the woods on that property, organised by the farm's cowboy, so much so that the park where the original event is currently held is named after this cowboy. As he explained, the name of the vaquejada park is "[...] Nilo Santos, this Nilo... The vaquejada in Porto da Folha, this cattle drive in the bush, started on Ilha do Ouro, with Nilo, it started in that dry land up there, in those highlands up there, that Nilo was the one who organised[11] ".

In this way, and in other ways, the process of social development of the riverside people of Porto-Folhense was presented, which was able to imply improvements in their affective attachments that still show traces of the customs lived by their ancestors.

[11] The source has not been revealed for reasons of preservation of rights.

CHAPTER 3

SOCIO-ENVIRONMENTAL IMPACTS

This chapter builds on the concept of the environment, socio-environmental impacts and points out the ubiquity between the environment and nature. It contextualises the environmental impacts due to the extinction of the flood cycle that existed naturally in the São Francisco River when it was not significantly damaged by human intervention. It also contextualises the social impacts that the Porto-Folhense riverside community of the Lower São Francisco suffered due to the absence of the water surplus responsible for the subsistence rice farming cycle.

The wording of this part draws attention to the contemporary world's concerns about man's actions on nature. These concerns have been intensely demonstrated in news reports in both the written and spoken press, as well as being brought up by academia, with regard to global warming, climate change, in short, care for the environment. In this case, it is understood that this human action is the realisation of the entrepreneurial event of the Xingó Hydroelectric Power Plant in the middle of the São Francisco River, because the episode happened by the force of man himself. The way everything happened, in this respect, on the environment uninhabited by man, the full flow (abundant river) which is a natural domain, shows that this space (environment) is understood to be the nature on which the anthropic action took place.

Implicit here is the ideological vision in search of sustainable development, both in terms of avoiding the dreaded global warming, and in the orientation that the transposition of this river will further aggravate the impacts identified. Not to mention that once this transposition has been carried out, its course is vulnerable to new impacts from the point interfered with by the channel that diverts it to its mouth. Although the terrestrial spheres, hydrosphere and atmosphere, are distinct, they promote a common factor, life both in and out of water, albeit in different ways. So, just as nature acts mysteriously in favour of life, when man interferes in his environment, his mysterious action or reaction is certain, thus leaving man subject to uncertainty.

That's why, according to Sachs (1998, p.163):

> Realising the ecological crisis forces us to think in long-term, global terms, transcending national borders. This observation remains valid even if political realism leads us to think that, over the next few decades, nation states will still have a decisive role to play, acting as protective shields against the negative effects of a globalisation process that is still insufficiently controlled through international agreements. In any case, the problem of managing humanity's common heritage - both natural and cultural - is currently emerging as a major challenge.

Having said that, the subject of this study raises the issue of socio-environmental impact. This impact can be understood as a consequence of certain causes engendered by human action on the environment. According to Almeida and Rigolin (2002), there is a harmony in the relationship between living beings and the environment, the so-called "ecological balance". When this is broken through human action, an environmental impact occurs. At first, this affects the physical environment and, from there, the problem can lead to other possible consequences for society.

The literature seen so far brings up a very complex variable: the environment. Alongside this, there are the resulting environmental and social impacts, which become perceptible in the environment as a result of human action.

As the socio-environmental impacts on the banks of the São Francisco River have affected the environment, which encompasses both nature and society, it is important to understand what the environment is and what environmental impact is.

The environment is what surrounds living beings and encompasses everything around them. In addition, its concept is broad and can encompass both what refers to man as social, as well as the space that encompasses said environment and nature. This environment is subject to the consequences caused by both society and nature.

In view of this, it is essential that man's relationship with nature is realised with responsibility, awareness and respect for both the environment and future generations. With regard to this, Carvalho (2003, p. 62):

> For example, something natural, like a river, doesn't necessarily have to be a source of energy, or serve as a sewer for industrial waste, but it could just be a watercourse from which someone draws food for survival, as the Indians do; everything will depend on the ideas we put into practice, which in turn will depend on the needs dictated by the lifestyle we choose to live.

Since beings are part of the environment, they can be present in both components that make up that environment, that is, if nature and society are taken into account as such components. Man, a rational being, is part of society, although his spatial organisation is contained in nature. Wild animals, irrational beings, belong to nature. Once domesticated, many of these animals can live together in society, even if the environment in which they live is artificial in relation to their natural habitat. This is just one of several examples that can help us understand the broad scope that the environment can encompass. In other words, the environment is not just restricted to the social environment or nature, but encompasses both. In addition to belonging to the environment and the environment belonging to nature, they are also fundamental factors in the existence of the environment.

According to Sánchez (2006, p. 18, 19):

> The concept of 'environment' in the field of environmental planning and management is broad, multifaceted and malleable. Broad because it can include both nature and society. Multifaceted because it can be understood from different perspectives. Malleable because, being broad and multifaceted, it can be reduced or expanded according to the needs of the analyst or the interests of those involved. [...].
>
> In this sense, the legal interpretation of the concept of 'environment' is decisive in defining the scope of environmental planning and management instruments. In many jurisdictions, environmental impact studies are not, in practice, limited to the physical and ecological repercussions of development projects, but also include their economic, social and cultural consequences.

According to Ross (2005, p. 198):

> In biology, understanding the environment means studying the structure and function of the different compartments of a system of interrelationships. In geography, in addition to this aspect, the environment is considered in its historical dimension and is directly related to human societies, in other words, the environment is nature transformed historically by culture.

As for Trepl (2006), the environment is ambiguous and is not seen in itself as the real existence of an environment. In other words, in the author's view, as this expression is understood today, it didn't exist two hundred years ago, even though almost everything that is understood today already existed. Finally, the author considers that in ecology, through the "organicist" system, communities or ecosystems self-organise, thus taking the liberty to constitute the environment. As for , in a second conception, especially ecological, according to the author, communities or ecosystems do not take liberties with their environment, only individual organisms have the character of self-organisation. "Ecosystems or communities of living beings are nothing more than abstractions for scientists, because it is the observer who **draws** their boundaries, and always differently, depending on the point of view." (TREPL, 2006, p. 349, emphasis added).

With all the above in mind, it is now time to understand the aforementioned parallel results for the environment, which are environmental and social impacts.

The idea of environmental impact can thus be traced back to the following thinkers:

According to Sánchez (2006), it is the modification of environmental virtue (in the sense of value) resulting from the metamorphosis of both natural and social processes due to human action.

Sene and Moreira (2000, p. 376) observe that "Environmental impact should be understood as an imbalance caused by a shock, an 'ecological trauma', resulting from man's action on the environment". In other words, it is the imbalance resulting from damage caused by man's action on the environment.

According to the Resolution of the National Environmental Council (CONAMA) - Brazilian Institute of the Environment and Renewable Natural Resources (IBAMA), No. 1/86, article 1, environmental impact is:

> Any alteration to the physical, chemical or biological properties of the environment, caused by any form of matter or energy resulting from human activities, which directly or indirectly affects: I - the health, safety and well-being of the population;
> II - social and economic activities;
> III - the aesthetic and sanitary conditions of the environment;
> IV - the quality of environmental resources.

This study emphasises the period that followed the 1920s, from the time when the rice-growing cycle in Porto da Folha/SE lasted until the present day. During this period, the region began to lose its spatial characteristics, in the socio-environmental and cultural sense, with the installation of various dams on the São Francisco riverbed. Despite this, this space suffered more significant losses after the foundation and full activation of the Xingó Hydroelectric Dam, from the mid-1980s to the beginning of the 21st century, 2004.

In light of the above concepts, it is also possible to understand that the problem in question can affect the local, regional and even the whole (understood as global or not), depending on the linkages of its nature. For example, the relationship between the bed of a river and its catchment area is direct, since if this link is broken, the effects may be noticeable from a given location to its region (the range that covers that location and its surroundings in the catchment area).

The impacts studied here have been unleashed on the environment, of which the study area is a part, which, in terms of scale, can be understood as follows.

According to Oliveira (1983), a locality is a place with or without inhabitants, smaller than a village, while a place is something that occupies a point on the earth's surface, i.e. a geo-referenced point in space, but with an administrative boundary, unlike the local attribute, which does not have this category. In terms of distinction, the local is understood to be contained within the place. Likewise, the concept of localisation of the area studied here is taken into account in terms of scale, in the sense of place, location and region. In other words, it is understood that there would be no region without the place and the whole would be empty or non-existent without its parts.

The socio-environmental impacts that have been unleashed on the banks of the São Francisco River have reached a certain point, a place, starting with the Ilha do Ouro floodplain, a place that has suffered significant consequences directly. This can be understood as a cross-section of space, the municipality of Porto da Folha/SE. In this conception, in the sense that this spatial section is part of the state of Sergipe, one of the cells of the entire Brazilian territory, to which the municipality belongs, the idea of a region can therefore be observed. This idea may be implied in relation to the location of the municipality and the Xingó Dam.

From this situation, it is possible to see the importance of studying regional geography as a means of advancing knowledge of the place studied. Otherwise, it would be possible to understand that science would be falling into impoverishment and would not be advancing towards its potential, new discoveries, which is common. As Lencioni observed (2003, p. 204):

> Despite these difficulties, or rather the loss of importance of Regional Geography, regional study is one of the most important fields for the advancement of knowledge, which, despite having been disputed by the most diverse disciplines, has found recognition by being included in geographical studies. So much so that there is no place, school or university, where there is a geographical concern, where there are no disciplines focused on regional study, even as an interdisciplinary complement. Here we are defending the position that, although the theme of the region can be interdisciplinary, Regional Geography is a particular disciplinary field within Geography.

It is also important to emphasise that this study discusses territoriality in the socio-environmental context. According to Ab'Sáber (2006, p. 30, emphasis added), "For numerous reasons, the concept of **total space** becomes the centre of consideration for correctly **predicting** impacts in the face of any type of project to be introduced in any area of a given territory".

As a result, the premise is that environmental impacts can result in social impacts, because according to Gonçalves (1992), among other authors, there are social and environmental environments. In this way, they are susceptible to impacts such as those that have occurred on the banks of the River São Francisco, in the Ilha do Ouro floodplain, in Porto da Folha/SE.

The starting point for the impacts studied here was anthropogenic action at first, i.e. when the hydroelectric project in question was set up. In a second moment, they were derived from this action, although they became evident in the area from a natural phenomenon, the flood cycle, when their extinction was confirmed. This was due to human interference when the Xingó Hydroelectric Dam was built on the São Francisco riverbed. Consequently, the extinction of the flood cycle triggered a series of other impacts on both nature and society.

The absence of surplus water on the Ilha do Ouro floodplain led to the suppression of the tropical hydrophilous forest and the water outcrops that existed there. As Rebouças (1999, p. 09) rightly observed:

> A portion of this surplus water forms the surface runoff that flows into natural rivers and lakes, causing local and occasional floods. [...].
> Another portion infiltrates the land in the catchment area in question, feeding the soil moisture, [...]. Soil moisture is the fundamental support that guarantees the development of an exuberant natural or cultivated plant biomass in the humid intertropics. As it is consumed where the rains occur, soil moisture constitutes localised water reserves.

The environmental impacts unleashed on nature have led to social impacts affecting the riverside community of Porto da Folha/SE. In other words, according to Silva (2002), a reservoir has the potential to transform both spatially and temporally the existing phenomena in a given watercourse. For this reason, it was also possible to observe that the extinction of the flood cycle caused the extinction of the local rice cultivation cycle. One of the main impacts on local society was the extinction of this type of culture. The way in which this cultural activity was carried out in the area could be understood as subsistence, since the sharecroppers cultivated the cereal for themselves to survive the whole year until they resumed this activity again. This depended on the return of the floods, which were regular and cyclical.

It is important to state that socio-environmental impact is the consequence caused by human action on the environment and that, as a result, the social environment is also affected, regardless of which level of the earth's spheres (atmosphere, hydrosphere and lithosphere) is affected, but never by natural causes. This is why, according to Santos (2009, p. 89), "[...] there will only be a socio-environmental impact if a community or society suffers consequences as a result of an environmental impact".

Therefore, when it comes to drawing up technical and scientific work on the implementation of a project in a particular region, these results must be foreseen before they are unleashed on the environment. Therefore, according to Ab'Sáber (2006, p. 29), it is "necessary to bear in mind the pre-existing scenario of spatial occupation and environmental quality, as well as the possible scenarios to be created at different depths in future times".

It is hoped that the new socio-spatial configurations will not be vulnerable to the consequences (socio-environmental impacts) of those who crave economic development at all costs.

In short, as Tropmair (1985, p. 01) rightly observed:

> Let's remember that the socio-economic and environmental problems that afflict us will not be solved by mere speculation: it will take a policy geared towards the real interests of communities, freedom and serenity for a wide-ranging debate and, in the search for solutions, solidarity will have to overcome selfishness and indifference.

ENVIRONMENTAL IMPACTS: EXTINCTION OF THE FLOOD CYCLE OF THE SÃO FRANCISCO RIVER AND OTHER CONSEQUENCES

After analysing the concepts of the various authors on the environmental issue, it is possible to state: environmental impact is the imbalance

This is the result of damage caused by various agents capable of disrupting the harmony that exists in the relationship between living beings and nature as a result of man's actions on the environment.

In this respect, according to Almeida and Rigolin (2002, p. 159) "We can say that environmental impacts are a kind of 'shock' that disrupts the ecological balance".

Through the concept of nature, environment, environmental impact and social impact, it is possible to understand how the impacts triggered by the existence of the Xingó Dam have affected the place, 80 kilometres downstream from the dam, on the bank of the river in question.

The dam was built at the end of the 20th century, in the 1990s, interfering with the bed of the São Francisco River, which bathes the river plains of the area studied. When the riverbed was dammed, everything around it and downstream was affected. The São Francisco River's flood cycle was totally dependent on natural flow. So, as the river began to suffer interference, the impacts began to unfold.

It can be seen that the cause of the extinction of the São Francisco River flood cycle, in Porto da Folha/SE, came from the control of its flow, because in this river basin, according to Santos and Câmara (2002, p. 82):

> [...] the projections of water demand for irrigation, for transposition to other hydrographic basins and for maintenance of the current hydroelectric dams raise concerns. According to data from the Secretariat of Water Resources of the Ministry of the Environment - SRH (1999), the total demand for water use licences in this basin is around 770m^3/s, of which around 99% is expected to be for irrigation projects. With a total average flow at the mouth of the river of approximately 2,850m^3/s, the demanded flow corresponds to 27% of the total flow. It is therefore imperative that more careful studies be carried out for licence applications, since the impacts generated could cause great harm to society.

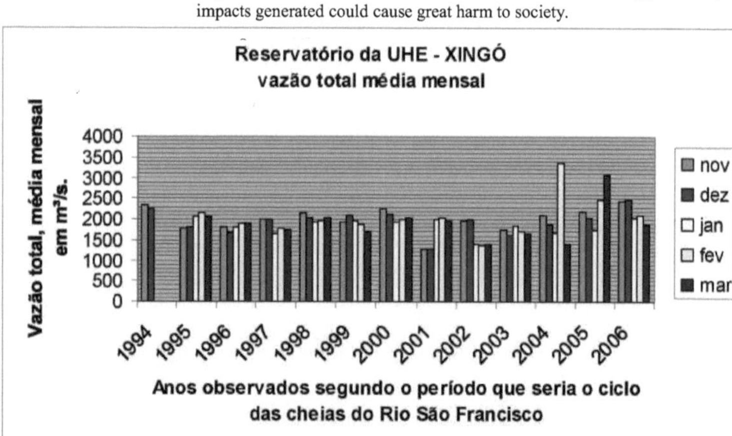

Graph 2. Average monthly flow in the XINGÓ HPP reservoir (1994/2006), with reference to the months of November/December and January/March. Source: Chesf - 2007.

By analysing the graph (no. 2), it is possible to see that the control of the river's flow was one of the causes of the extinction of its natural flood cycle. Taking into account the years 1994/2003 shown in the graph below, the river should have started flooding during the months shown, according to Vargas (1999). However, at no point during these years was the river's total average flow, 2,850 m^3/s, reached.

Despite this, it was noted that in 2004 and 2005, in the months of February and March respectively, the flow was adequate for the occurrence of floods. These floods were dangerous because they hit the riverside community unexpectedly. Thus, causing immeasurable, irreversible and even fatal impacts.

If the Xingó Dam, according to the graph (no. 2) above, did not release a flow of 2,850 m3/s between 1994/2003, it is not to be expected that the São Francisco River will reach this volume at the mouth. Especially since a few kilometres upstream from the mouth, there has already been talk of saltwater fish being caught, not to mention other impacts that have also been claimed due to the drowning of the mouth. In this case, it was the action of the sea at this location, so that due to the very low volume of the river's flow, part of the mouth area was submerged by the sea. As a result, Vila Cabeço, which was located at the mouth, disappeared.

Figure 12. Mouth of the São Francisco River, border between Alagoas and Sergipe. Source: author's photos, July 2008.

You can see in the picture (no. 12) above that only the ruin of the lighthouse remains, where the village was located around it. To further clarify this reasoning, according to Silva (2002, p. 36) "Fluvial erosion is changing the riverbed, while coastal erosion is causing the sea to advance over areas located on the coast".

At the top of the picture (no. 12), you can see, between the arrows, the mouth of the River São Francisco in an estuary about 2 kilometres long: on the right-hand side, it shows vegetation from the territory of Sergipe and, on the left-hand side, part of the territory of Alagoas. Below, on the left,

In the foreground, you can see the action of the sea, which is responsible for the erosion that can be seen there, so much so that the roots of the coconut trees are exposed and they are leaning. In the background, among the remains of the coconut trees, you can see the lighthouse where the village of Vila Cabeço once stood. On the right, it's clear that the Atlantic Ocean, with its waves, is more powerful than the São Francisco River. Still in this position, in the background is the aforementioned lighthouse.

This river crosses a large expanse of semi-arid climate, from the Xingó Dam to its outflow into the sea, so it is unlikely to receive enough water to reach a flow of 2,850 m^3/s. This is because in dry periods, the São Francisco can contribute from the mouths of its ephemeral tributaries, which exist from the dam to its mouth in the Atlantic. In this respect, according to Rebouças (1999, p. 09, 11), "[...] the portion that infiltrates and feeds the groundwater flows constitutes a mobile reserve, as it can be captured outside the area [...] and will supply the rivers during periods of drought".

On the one hand, there is a notorious difference between the flow at a certain point along the river, considering the Xingó Dam, and the next point reached, for example, its mouth. In other words, before the existence of the dams on the São Francisco riverbed, during the flood period its flow rate rose "to 13,000 m3 of water per second, and during the ebb, when this same rate becomes less than 900 m3 [per second]." (ANDRADE, 2004, p. 140).

On the other hand, according to the company contracted by Companhia Hidro Elétrica do São Francisco (Chesf), called Engenharia e Consultoria S.A. (Enge-RIO) (1993c, p. 166):

a) The low flow period is from June to November, while the highest flows occur from January to May;

b) the natural monthly average flow for this period is 2,980 m^3/s;

Analysing the graph (no. 3) below, it can be seen that the Xingó Dam's outflow only reached the level of the São Francisco's average flow for two months, as mentioned in its Environmental Impact Study and Environmental Impact Report (EIA/Rima). The period observed was the only time when the peak flow was recorded, from when the Xingó reservoir came into existence until the date this flow was recorded.

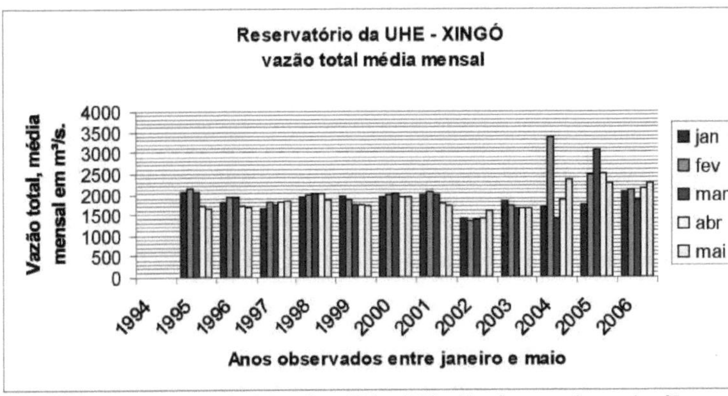

Graph 3. Flow recorded at the Xingó Dam from 1994 to 2006, with reference to the months of January to May, the only period to record a peak flow. Source: Chesf - 2007.

There was a time when the river could have reached a flow of 2,850 m^3/s, which was expected to be an adequate minimum level to exceed on the São Francisco river plains. This would be possible if, at the very least, the river's volume were already close to normal at that time and if it were in its present abundant flow, as it always was. This would consequently result in floods from the initial months, November/December, according to an analysis of the data and information in this work. However, according to Graph 2, for the years 1994 to 2003, the volume of this flow was not reached.

According to the graph (no. 4) below, the outflow from the Sobradinho reservoir from 1979 to 1986 was higher than the flow of 2,850 m^3/s in most of the months observed. According to Vargas (1999), the months from November to March are the reference period for the São Francisco River's flood cycle. On the other hand, when the civil works for the Xingó hydroelectric project began in 1987, it is possible to see a decrease in the flow.

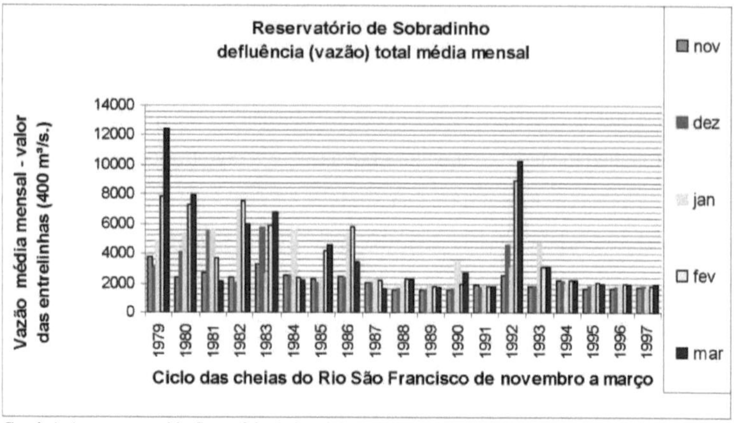

Graph 4. Average monthly flow of the Sobradinho reservoir 1979/1997. Source: Chesf - 2007.

As the dams were installed on the riverbed, the flood cycle was de-characterised until it was interrupted by the Xingó Hydroelectric Plant (UHE-Xingó). This is so evident that Graph 4 shows that 1995 saw a peak average monthly flow in February of 2,025 m^3/s, lower than the 2,850 m3/s flow. However, that flow was recorded at the Sobradinho reservoir and not at the mouth of the river. Now, looking at the graph (no. 5) below, the above reasoning is confirmed, as the Itaparica reservoir, downstream of Sobradinho, recorded a flow of 2,118 m3/s in February, despite the fact that its inflow was 2,146 m3/s.

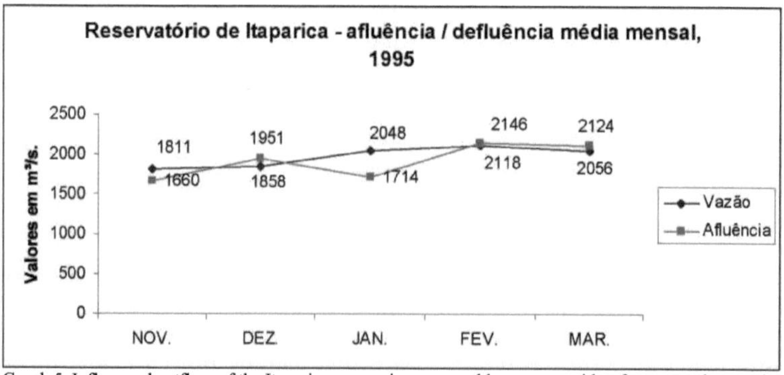

Graph 5. Inflow and outflow of the Itaparica reservoir as a monthly average with reference to the months of November/December and January to March 1995. Source: Chesf - 2007.

These figures are higher than the flow released by Sobradinho, but lower than 2,850 m^3/s, although due to the great distance of this reservoir from Porto da Folha/SE, the floods have been adding up along the river until they reach the Ilha do Ouro floodplain. We can see that this is because until the river bed reaches this location, it receives several tributaries that could contribute to the increase in the São Francisco's flow. The opposite occurs from the Xingó Dam to the study area in relation to the hydrographic network (tributaries of the São Francisco). Finally, according to Sudene/Adene (2007), in 1995 the São Francisco River flooded the municipality of Porto da Folha/SE.

According to Silva (2002, p. 29), "Regardless of the size of the reservoir or the purpose of the water accumulated in it, its main function is as a regulator, with the aim of maintaining the flow of watercourses or meeting variations in user demand".

It was precisely because of the regulation of the flow in the dams that the flood cycle was extinguished, so much so that today the total average monthly defluence recorded in the Xingó reservoir has reached 2400m3/s. It is clear that this figure is not the one recorded at the mouth of the river, because if

If this were the case, the flow would be much lower because of the semi-arid region that the river runs through from the dam, as well as the water difficulties, also due to the climate, that the region faces from its various ephemeral tributaries. Even the action of the sea is already interfering with the riverbed. Its flow is so reduced compared to the flood cycle that its strength no longer exceeds that of the sea near its mouth.

It is important to emphasise that, without the dams in its bed, the São Francisco River flowed as time went by, according to the seasons, its level following the dynamics of nature. The period when the floods hit the floodplains of Porto da Folha/SE was exactly when the river had already met its needs downstream, all the way to its mouth. In this way, its flow had the force to push the river sediments into the Atlantic Ocean.

When the river had already received enough water, its level was normal, with a large volume. The natural and frequent end-of-year rains common in the Southeast, in Minas Gerais, contributed to the increase in the flow of the São Francisco, creating the flood cycle that existed for more than a century in Porto da Folha/SE. At that time, between the middle and end of the year, the river's water level remained very high for a long time, which led to the occurrence of the flood cycle in the area. As a result, the excess water coming from upstream was added to the localised levels in the region.

The cyclical period in which the São Francisco River flooded the area is shown in graph 6. Most of the interviewees confirmed that the Ilha do Ouro floodplain was inundated from November to March. Even so, the water remained over the Várzea until the beginning of April, when the rice crop grew.

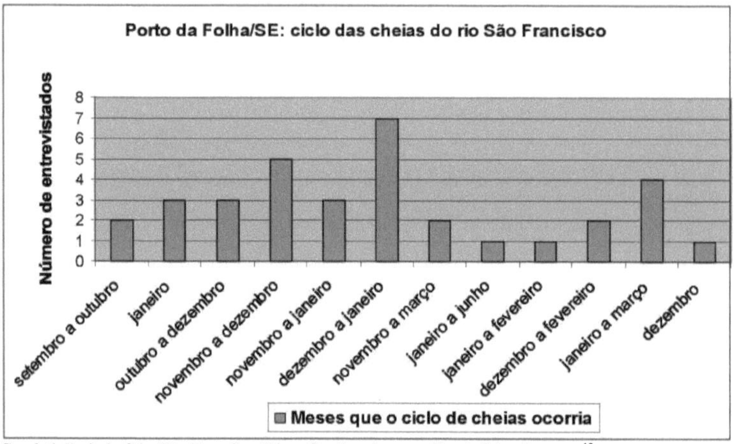

Graph 6. Period of the São Francisco River flood cycle according to interviewees .[12]

According to Santos and Câmara (2002, p. 155):

> Gradual flooding is characteristic of large river basins and lowland rivers such as the Amazon, the Paraguay, the Nile and the Mississippi-Missouri. The phenomenon evolves in an easily predictable way and the flood wave develops slowly from upstream to downstream, at regular intervals. The water rises in flood for a while and then gradually drains away. Gradual flooding is usually cyclical and clearly seasonal.

This concept corroborates the process of the flood cycle that affected the main floodplain and the entire river plain area of Porto da Folha/SE. Despite being a plateau river, the same natural process occurred with the São Francisco River. The flood cycle of this river that reached the town was compared to that of those rivers at due to the great length that its bed travelled freely from its source until it reached Porto da Folha/SE. However, the opposite happened with the Capivara River. When it floods torrentially, its waters come and devastate everything, so much so that you can hear the roar of the torrent taking whatever is in the dry bed,

[12] Source: interviews conducted by Regnaldo Gouveia dos Santos in January 2007.

such as the animals and whatever is in front of it.

With the existence of dams on the São Francisco riverbed, especially after the Xingó works, the dam had to be filled in as time went by in order to become active. After this period, the river's flow continued, but it was controlled by this dam, which is much closer to the site studied than the Moxotó reservoir/AL and the Paulo Afonso hydroelectric complex/BA.

It was possible to observe that over the course of the year, the defluence, the release of water, from Xingó in several years has not been able to meet the needs of the flow of the São Francisco all the way to its mouth.

The more the outflows began to occur in a controlled manner, over the course of each year and at each moment, the less the river (nature) found its fullness (preservation of the vegetation on its banks, determination of small islands in its bed, etc.). This unfulfilled fullness refers to the river's encounter with the lush vegetation that existed around the Ilha do Ouro floodplain until the late 1970s and mid-1980s.

This is why the water surplus no longer exists on the river plains of Porto da Folha/SE, since the effects caused by the Xingó Dam are directly related to controlling the flow of the São Francisco. Since the existence of the various dams on the riverbed, especially the one mentioned above, has led to the regularisation of the flow in the area. Finally, according to the graphs above, this has resulted in the extinction of the flood cycle.

The place, which encompasses the entire river plain as well as the main floodplain, on the banks of the Capivara that surrounds the municipal centre, when the São Francisco River floods with its certain frequency, implied an enclave in the sertão of Sergipe. Taking into account all the lush forest that bordered the floodplain known as Ilha do Ouro and the town's surroundings.

In other words, the enclave was confirmed because the water there was abundant, contributing to the local water table, which supplied the needs of the vegetation around the floodplains and the municipal centre; it also contributed with water outcrops. As a result, for the duration of the flood cycle, the place was like an island in the sertão of Sergipe, surrounded by the caatinga and the difficulties of the Brazilian semi-arid region faced by the region's inhabitants. In this sense, it is possible to observe that the saturation of the water table determined the existence of the tropical hydrophilous floodplain forest, which existed from the foot of the slopes and extended to an altitude of around 100 m (one hundred metres) on the banks of the Capivara in the circumscription of the place, on the bank of the São Francisco.

The existence of this forest depended on the humidity of the soil, which is fundamental for the development of the exuberant natural plant biomass. Rebouças (1999, p. 117) observed, "In fact, a fraction of the water that infiltrates the surface of submerged lands will constitute soil moisture, which is the fundamental support for the Earth's plant biomass and the atmosphere/lithosphere interface". This concept corroborates the enclave rationale. As a result of this biomass, the fauna and ecosystems of the place rhythmised themselves to the existence of that enclave. This is why the surplus of water that reached the Várzea was decisive in the conservation of the hydrophilous tropical forest and in the ecosystem balance of the place.

Figure 13. Water outcrop on the Lagoa Comprida road, Porto da Folha/SE. Source: photo by the author, December 2006.

Water outcrops were checked according to the sign in figure 13, at the foot of the Serra da Lagoa Comprida, under a tree native to the banks of the São Francisco River, a craibeira (bignoniaceae). These points were noticeable to anyone who passed by, because "when it rained and when the floodplains were full, it mined on the Lagoa Comprida road, near Miguel Chorão and Seu Pedrinho da Olaria, under the Craibeiras" (verbal information) .[13]

In this way, it was possible to observe that, with the leaching process of that mountain range, in the rainy season from February, March to June and July, coinciding with the flood cycle, the small mines would emerge in the form of puddles, because, according to Rebouças (1999, p. 123):

> [...] underground flows in aquifers, whether free or confined, follow more or less long trajectories, heading from the sectors with the highest hydraulic potentials to those that are comparatively lower. In the specific conditions of a free aquifer, this means that the flows take place from the hill sectors towards the valleys.

The large tree on the left-hand side of the picture (no. 13) above is an old craibeira that withstood the impacts. You can still see traces of the points where the water naturally rose at the foot of the mountain, in the middle of the Lagoa Comprida road, as shown. A few metres to the right, the land falls to the bed of the meandering Capivara River, which is located on the extensive river plain of Porto da Folha/SE.

Through pools, the water came to the surface, but only during the cyclical flooding period, when the water table was dense due to the flooding of the São Francisco that occurred over the floodplains. These points were localised, i.e. they weren't scattered all over the place, according to Santos and Câmara (2002, p. 88):

> In turn, the water flow system in the subsoil of the hydrographic basin in question may cover local, intermediate or regional dimensions, in which the groundwater transit times may comprise days, years, centuries or millennia respectively [...].

The lack of surplus water has caused immeasurable effects on the floodplains of the municipality of Porto da Folha/SE and the extent of these on the banks of the Capivara Creek. As the Xingó hydroelectric project was installed, causing interference with the riverbed by damming its waters, the environmental impacts became apparent in the place studied. It was the flood cycle that determined the existence of the tropical hydrophilous floodplain forest.

Environmental impacts were unleashed in the area in a way that altered the physical environment, causing an imbalance in the local ecosystem. This imbalance was evident as the natural floods of the São Francisco stopped being

[13] The source has not been revealed for reasons of preservation of rights.

cyclical. As a result, the tropical hydrophilous floodplain forest began to lose its lush green colour due to the lack of surplus water around it. As a result, some species of birds, such as the magpie and the sofrê, among others, which used this area as their natural habitat, became practically extinct and have not been seen for a long time. These impacts covered the entire river plain along the banks of the Capivara River, which surrounds the municipal centre. As a result, a number of impacts became apparent, including the lowering of the water table (the place that contains water in the shallow part of the subsoil). This was a determining factor in the existence of the tropical hydrophilous floodplain forest and the water outcrops. Thus, the impact on the water table led to the extinction of the lush vegetation and water outcrops.

This happened because of human activity when the Xingó Hydroelectric Dam was built. If you compare the place today to the rice-growing period, you'll see that it's deserted, as opposed to the contingent of farmers growing rice in the Várzea at that time.

In the area studied, there is an extreme water deficiency in relation to the São Francisco flood cycle, during the same period in which this natural phenomenon could be observed. According to Conti (1998), the area is showing signs of desertification.

Figures 14 and 15 clarify this reasoning about environmental impacts, as the map (fig. 14) shows:

a) In the foreground, you can see the flooded area shown on the map (fig. 14), which was responsible for rice production and the presence of the water table. This, in turn, engendered the lush forest between 21 and 100 metres above sea level, which bordered the flooded area.

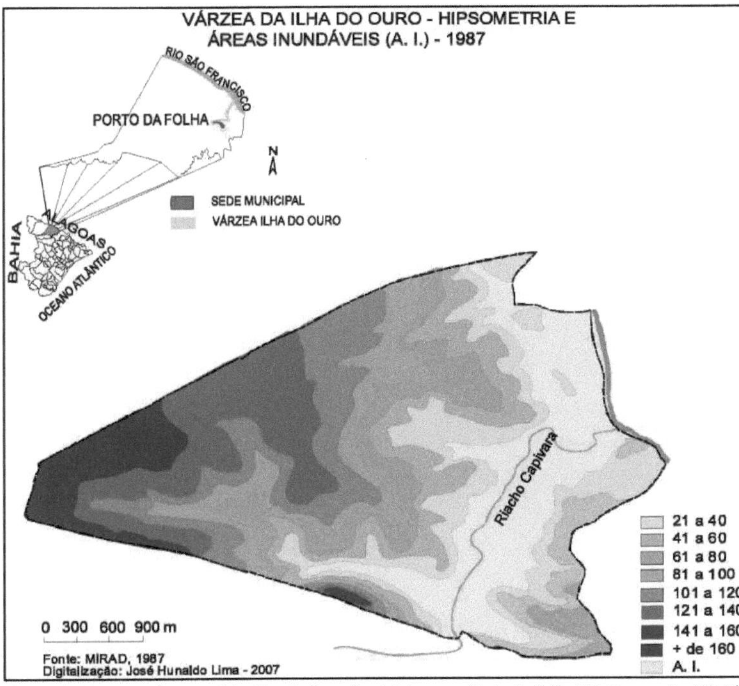

Figure 14. Hypsometric map of the Ilha do Ouro floodplain area - Porto da Folha/SE. Source: José H. Lima - 2007.

b) From an altitude of 120 metres, you could see the caatinga, the characteristic semi-arid vegetation.

If you look at the picture (no. 15) below, from the same area, the effect of the extinction of the São Francisco flood cycle is notorious, as you can see both the absence of the water surplus in question and the exuberant tropical hydrophilous floodplain forest. According to this figure (no. 15), in the background you can see the flow of the São Francisco. Embedded in the valley, you can see the land of whose floodplain this work contextualises the socio-environmental impacts on the banks of the river. Also visible are the meanders of the Capivara River towards its mouth and, on the slopes of the valley, the characteristic caatinga of the northeastern semi-arid region.

Figure 15. Ilha do Ouro floodplain at the time of the São Francisco River's flood cycle. Source: photo by the author, December 2006.

This landscape would not have been possible to perceive at the time of year when the photograph was taken if it had been during the emblematic flood cycle. In other words, if this photograph had been taken in the period before the hydroelectric power plants were built, especially the Xingo plant, the Várzea would have been flooded and all the vegetation, as well as being dense, would have been green.

In view of this, it is important to emphasise how Ibge (1994, p. 101) observed the environmental problem in the São Francisco Depression:

> In the floodplains and terraces of the São Francisco, which are highly unstable because they are subject to fluvial dynamics, the environmental repercussions are reflected in the truncation of the topsoil, furrowing, compaction and turbidity of the water. In these environments, the current situation reflects a strong compromise accumulated over time, resulting in a state of environmental quality classified as Degraded.

With reference to this study carried out by Ibge (1994), one can observe that the environmental problem in the São Francisco Depression revolves around the droughts and floods that hit the São Francisco River basin from the 16th century to the second half of the 20th century. This study refers to floodplains as being highly unstable because they are exposed to fluvial dynamics. This conception is consistent with the present study because the cyclical natural floods of the São Francisco brought great socio-environmental benefits and, as Vargas

(1999) rightly observed, the frequent floods in the lower reaches of the river had this natural function. Natural, because the sediments that the river deposited on the floodplains, by force of nature, were true natural fertilisers. The social benefit was confirmed in the good rice harvests that determined the livelihood of the riverside community, which also depended on the natural fertility of the floodplains.

As a result of the extinction of the flood cycle, some signs of the desertification process have emerged on the Ilha do Ouro floodplain, such as the presence of saltpetre due to the flooding of the Capivara Creek, complained of by the interviewees.

The figure (no. 16) below corroborates the above assertion, as it is possible to see concentrated salinisation at some points, among other signs of this process such as the lack of regenerative capacity of the soil, at least at these points. In addition, the cracked soil shows the environmental impact such as ecological imbalance due to the lack of surplus water.

Among other indicators, it is also possible to note the absence of the large contingent of people who used to cultivate rice in the Várzea, after the extinction of the rice cultivation cycle and the suppression of the Várzea forest. As a result, today there are only a few trees left, those that resisted the impacts. Not to mention another very important indicator, the reduction in biodiversity. In this respect, both the cactus deer and the fish in the floodplain lagoons have disappeared due to the lack of surplus water.

Figure 16. Aspects of the Ilha do Ouro floodplain after the extinction of the São Francisco's natural flood cycle. Source: author's photos, December 2007 and January 2008.

The significant decrease in the fish population in the São Francisco riverbed is due to the following fact: analysing the EIA/Rima for the Xingó HPP, it was noted that, according to Enge-RIO (1993f), this document foresaw significant impacts on water quality.

Now, it is important to emphasise that in contrast to what Ibge (1994) observed, in the floodplains of Porto da Folha/SE, river dynamics were not seen as a problem, but as an environmental solution.

It should therefore be stressed that the floods of the São Francisco contributed to the natural ecosystem by depositing sediments and natural fertilisers in the lagoons of the flooded areas, as Vargas rightly observed

(1999).

Even so, in relation to the degradation of the floodplains, according to Ibge (1994), it would be possible to agree if it referred to the absence of river floods as the problem, because in this way it is possible to permeate the concept of desertification. According to Conti (1998, p. 67) "Ecological desertification occurs when ecosystems lose their capacity to regenerate, with the rarefaction of fauna and a reduction in the area covered by vegetation, followed by soil impoverishment and salinisation."

It was also in this sense that the consequences of the implementation of the Xingó Dam hit Porto da Folha/SE. As a result, several species of fauna that were common in the area are no longer found. Such species as the red deer (Mazama Gouazoubira) may be extinct because it hasn't been seen there for decades, the spider shrimp (Macrobachium) is no longer heard of and the piranha (piraya) is little mentioned. Of the Várzea's hydrophilous rainforest, only a few trees remain. All these species were abundant during the flood cycle. The soil of the Várzea is vulnerable to salinisation due to the absence of water from the São Francisco, which is considered a freshwater river, so when it reached the area it revitalised the environment. In other words, among other ways, it revitalised the soil both through natural fertilisation and by eliminating the salt concentrated due to the flooding of the Capivara or the time without water.

With regard to this, currently "if the Capivara floods, it leaves saltpetre on the land, the land is lost that nowadays when the Capivara floods, people want to plant rice and they can't..., because of the saltpetre in the waters, which is strong. And the São Francisco River removes the saltpetre... so it stays..." (verbal information) .[14]

The driving force behind the effects, environmental impacts and, consequently, social impacts, was the control of the river's flow, mainly at the Xingó Hydroelectric Dam. Despite this, the other dams, unless already mentioned, also influenced the triggering of these impacts on the place, but indirectly, so much so that they were unable to extinguish the flood cycle. The extinction of this cycle only became apparent after the Xingó hydroelectric project was completed.

SOCIAL IMPACTS: EXTINCTION OF SUBSISTENCE FARMING

The impacts that affected the riverside community of Porto-Folhense are considered social impacts because of the way they unfolded. The more time passed, from the start of the Xingó Dam works, the less the local community grew rice, which was one of the main staples of their meals. As a result, many members of the community went through great difficulties, such as having to cut down on

[14] The source has not been revealed for reasons of preservation of rights.

the number of meals they ate, as production of the cereal was drastically reduced until it was extinguished after the hydroelectric dam came into operation. This can be seen in graph 7, because as each hydroelectric project was installed on the São Francisco riverbed, from Sobradinho to Xingó, the impacts added up. However, they only became noticeable in a significant way when the municipality's rice crop was extinguished after the Xingó Dam was built in the mid-1990s. It is important to emphasise that after the Paulo Afonso/BA complex was built, inaugurated in 1955, the graph shows a significant increase in rice production in Porto da Folha/SE, rather than a decrease. The 16 kilometres^2 of area flooded by the PA IV reservoir, the last plant installed in the hydroelectric complex, is the smallest of the many powerful hydroelectric plants in Brazil and one of the smallest in the world, also taking into account its hydroelectric potential compared to these same plants. That's why the Paulo Afonso complex is a world example of how to produce energy from hydropower. In other words, the production of hydroelectric power is socially, economically and environmentally viable, based on suitable geographical areas such as high waterfalls and where it is not necessary to dam up a lot of water for this purpose.

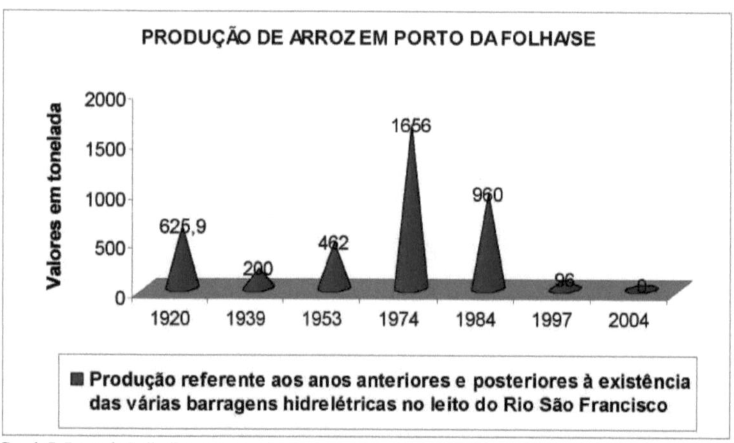

Graph 7. Porto da Folha/SE, comparative annual rice production. Source: IBGE, general census and municipal agricultural production for the respective years.

Although the figure shown in the graph (no. 7) for 1920 is not as high as those shown for 1974 and 1984, in that year Porto da Folha/SE had the third highest production in the state. Neither

The start of work on the Sobradinho/BA dam in 1973 had an impact on rice production in the area, which changed considerably after the last generator of the Xingó HPP went into operation in 1996.

The interviews corroborate the above assertions. This was particularly noticeable after the Xingó Dam was activated. Bastos and Fonseca (1997, p. 12) confirm the occurrence of this process:

> With the construction of dams to form hydroelectric lakes [...] throughout the São Francisco River basin, the river level has changed, with lower and lower levels in the portion downstream of Paulo Afonso. With these changes, the fish have 'disappeared' from the river [...]. The fishermen complain about the lack of conditions for them to continue to survive by fishing and the local population complains about the lack of this food, which has always been an alternative source of protein.

It is in this sense that concerns arose about the hydroelectric projects that began to be implemented, historically, in the countries of the South in the 70s. This is why Teixeira et al. (2006, p.
164), that "The 70s constitute the historical landmark of the emergence and expansion of these projects in the so-called Third World", although some hydroelectric plants appeared at the end of the 60s.

It was also in the 1970s that human concerns about environmental issues began to spread. According to Ab'Sáber (2006), at the Stockholm meeting in 1972, human intelligence reacted by introducing the concept of eco-development.

Around the second half of the 20th century, Brazilian environmentalists who were sensitive to the major development projects implemented in the country became concerned about the future of the people and their socio-environmental environment, when they opted to develop mechanisms to prevent impacts.

This concern can be seen in the following
Ab'Saber (2006, p. 35) states that, "Predicting impacts is an act of taking precautions to guarantee harmony and make functions compatible within the total space in the future".

One of the concerns for the future of mankind is preserving the environment in which life exists. Turning to geographical science is one of the ways to understand the causes and consequences of the environmental and social impacts unleashed on the banks of the São Francisco River. Firstly, because geography is one of the fundamental sciences for such a study due to its socio-spatial scope.

And secondly, studying environmental impacts is necessary because they engender social impacts.

It is important to emphasise that the social impacts of this problem, discussed here, afflicted the community of Porto da Folha/SE after the extinction of rice cultivation. This type of crop, which existed until the end of the 20th century, was a determining factor in the subsistence of Porto da Folha's riverside community. The sharecroppers exchanged part of the rice crop for other types of necessary food and the other part was kept for consumption throughout the year.

Thus, "In terms of the consumer market, it can be said that although rice is widely consumed among the lower income classes, it is also part of the diet of the majority of wealthier consumers [...]" (SILVA, 1978, p. 185). This type of food was destined for the tables of the social class far above the local population. Despite this, the lower classes regularly consumed the food in question without having to pay in kind, which is no longer the case today. This shows that there was a strong denial of contemporary social distinction, in the sense that nowadays it is necessary to have a high capital value in order to survive, whereas at that time the agricultural system provided that livelihood.

This crop was subsistence because it was for self-consumption, according to interviews with the target community.

The techniques used to manage the rice crop there revolved around the use of rudimentary tools such as the hoe, mattock, shovel, among others, as well as the sickle to carry out the swidden. As a result, this community acted like "Human groups, endowed with soft technologies, adapted to the conditions and rhythms of nature, were very little aggressive and comprehensive in the face of the different attributes of regional territoriality." (AB'SÁBER, 2006, p. 32). At that time, in Porto da Folha/SE, rice cultivation developed in this way.

Figure 17. Ilha do Ouro floodplain (cultivation of winter crops such as maize, beans, etc.). Source: photo by the author, July 2006.

Analysing figure 17 above, you can see the same descriptive content as figure 15, but with very similar characteristics to the time when the natural flood cycle existed, although the crop seen in the photo is not rice. At that time it was also possible to see maize and beans planted in association with the cereal, but limited to the combros. On the slopes towards the valley, the remnants of the dense plant biomass are noticeable, although the same plant layer that existed at the time of the natural flood cycle is not present.

It can be seen that the Xingó EIA/Rima did not consider an adequate area of real influence for the implementation of the project. Mainly because the case involves interference in the bed of a river (São Francisco), which, from the construction site onwards, runs for around 180 kilometres to its mouth. All the points that are connected to the riverbed along this stretch are vulnerable to the impacts of this interference, since the volume of its flow has been altered.

On the one hand, for Monosowski (2006, p. 139):

> A paradox therefore arises in the case of the environmental assessment of individual projects. If the analysis comes up with negative results, its implementation is

> formally discouraged. However, if we conclude that the assessment is positive, this does not mean that the project is necessarily appropriate in relation to global reality.

On the other hand, Alves (1995, p. 65, 66) observes that:

> The EIA is a technical-scientific analytical procedure carried out by a multidisciplinary team [...]. It will thus serve as an ecological planning aid to instruct the competent authority, which will have to grant the corresponding environmental licence on the basis of the information it contains.

Thus, the level of impact triggered by the realisation of large undertakings of the most diverse circumstances, such as the execution of hydroelectric projects, should, before starting their work,

be evaluated and discussed, both by scientists and technicians, as well as by the local and general public. Only then can the appropriate decisions be made about the implementation of the project. However, this decision must not lead to any short, medium or long-term social or environmental impacts on the area covered by these works. What's more, any and all regions directly or indirectly affected by the project should be the subject of discussion about the extent to which the project may or may not interfere.

With this in mind, the following fact is noteworthy: the EIA/Rima for the Xingó HPP did not consider the site studied here to be an area of influence for the project. This statement instigates the present work, opening up a relevant discussion on the case, since Porto da Folha/SE is on the banks of the São Francisco River and this, in turn, has suffered flow interference due to the existence of the project. This has resulted in environmental and social impacts on the area in the short and medium term. As a result, in this case, the EIA/Rima did not open up a real area of influence in relation to the development in question and the areas that are in direct contact with the interfered riverbed. In this respect, according to the National Environment Council - Ibama (1986), Conama Resolution No. 01/86, Article 5:

> The environmental impact study, in addition to complying with legislation, in particular the principles and objectives expressed in the National Environmental Policy Law, will comply with the following general guidelines: III - Define the limits of the geographical area to be directly or indirectly affected by the impacts, called the project's area of influence, considering, in all cases, the hydrographic basin in which it is located;

As Alves (1995) rightly pointed out, the information contained in the EIA shows the responsibility of those involved in drawing up a project of such socio-environmental value. According to

Ab'Sáber (2006, p. 28), "This is why the teams responsible for assessing impacts have a great cultural and moral responsibility, which can result in great success or a total loss of credibility in the eyes of their country's scientific community".

It is in this conception that the errors of human action are certainly denoted through their interference with the environment in which man himself lives.

In view of the immense impacts suffered by the area studied due to the extinction of the São Francisco River's flood cycle, it is possible to observe here a flaw on the part of those who conducted the study. In other words, the Xingó EIA/Rima did not consider the site studied here in relation to the dam as an area of direct influence, nor of indirect influence, since for Engenharia e Consultoria S.A. (Enge-RIO) (1993e, p. 01):

> The Area of Direct Influence (AID) was the area formed by the municipalities of Piranhas [AL] and Canindé de São Francisco [SE] and the Area of Indirect Influence (AII) was the area formed by the municipalities of Olho d'Água do Casado [AL], Delmiro Gouveia [AL] and Poço Redondo [SE].

The last place downstream of the dam considered by the EIA/Rima of the Xingó HPP as an area of indirect influence is the municipality of Poço Redondo. Although the region studied here is bathed by the river and is about 80 kilometres downstream from the dam's location, it was nevertheless affected by the sharp drop in the river's flow. Until the Xingó dams were built, the river flooded all the floodplains of Porto da Folha/SE and flowed into the mouth of the Capivara River, turning around at the municipal centre.

This is confirmed by the fact that Enge-RIO (1993e) did not consider the municipality of Porto da Folha/SE to be within the dam's area of influence, although Ibge at the time considered it to be within the region of Canindé de São Francisco/SE, where the Xingó HPP is located. In this respect, the Sergipe micro-region of the Sertão do São Francisco, and therefore Xingó's area of influence, is: "Canindé de São Francisco, Feira Nova, Gararu, Gracho Cardoso, Itabi, Monte Alegre de Sergipe, Nossa Senhora da Glória, Poço Redondo, Porto da Folha" (Enge-RIO, 1993e, p. 05).

There is no doubt that "A reservoir imposes significant changes in the temporal and spatial scale of the **phenomena** that occur in a watercourse."

(SILVA, 2002, p. 29, emphasis added).

Thus, the phenomena that occurred in the area, such as the end of the flood cycle and the drastic reduction in the fish population, can be understood as the possibilities that led to the ratification of the social impacts. These were confirmed because of anthropic action in carrying out a certain event, the execution of the Xingó hydroelectric project.

In this way, it is possible to see the influence that the Xingó Hydroelectric Dam had at the time its EIA/Rima was drawn up.

It is important to bear in mind that predicting impacts in relation to a project such as the one in question here, among others, according to

Ab'Sáber (2006, p. 27):

> [...] it is an essentially multidisciplinary technical-scientific operation of great importance for Third World countries. Firstly, because it reveals the level of enlightenment achieved by the country's society in terms of its ability to foresee future spatial organisation of its territory. And secondly, because it is also a good indicator of the strength of social pressure from enlightened groups in relation to the proper use of legal instruments to guarantee a reasonable framework of environmental quality and land-use planning beforehand. Finally, because it is an excellent test for evaluating the potential of available legislation, as well as its applicability to specific cases.

When the natural flood cycle of the São Francisco River occurred, the entire area of the river, including the Capivara Creek, was entirely flooded, forming a single river reaching the foothills of the slopes.

protect the city.

Through their mysticism, the local community expected the floods of the "Old Chico" to be on the right dates, because their lives depended on them. In this sense, Cunha (1997, p. 98) observed, "It is not only the knowledge (knowing) of the conditions of the place, but also the belief in the divine power, since for the sertanejo this imaginary, the semi-arid sertão, is also the result of God's creation".

Even in times of great localised droughts in the sertão, which lasted in the 1980s, 1980-1984, for four consecutive years, in addition to 1987-1988, that community resorted to the Ilha do Ouro floodplain as a means of refuge from hunger. This was only possible with the existence of the natural flood cycle of the miracle-working São Francisco River.

With regard to the influence of the Xingó Dam, the record of floods in the

Sudene/Adene department (personal information)[15] , which occurred in 1975, 1979 and 1985, is noteworthy. As a result, it was noted that the last two years recorded date back to the regulation of Sobradinho's flows, because according to Ibge (1994), this dam was not designed to control river floods, although it has had this function since 1979. In other words, even though the largest of the dams existed on the São Francisco riverbed and, above all, with its flood control function, the flood cycle persisted in the area studied. This cycle was only extinguished in the mid-1990s, with the end of work on the Xingó Dam, the closest to the site, because according to Enge-RIO (1993b, p. 40), "The sixth and final generating unit of stage 1ª will start operating in April 1996".

In view of this, it is worth emphasising that the extinction of the rice-growing cycle in Porto da Folha/SE is akin to the extinction of the Rio flood cycle

São Francisco as it is for the intervention of the riverbed through the dams installed there, especially the Xingó Hydroelectric Plant. According to Monteiro (1962, p. 14), "If between Piranhas and Pão de Açúcar they [the rice-growing floodplains] are sporadic, their frequency increases progressively from there downstream, so that they dominate the valley from Própria to the mouth".

In accordance with the author's meaning, it is also confirmed that the municipality of Porto da Folha/SE certainly has floodplains.

15 Sudene/Adene - Superintendence for the Development of the Northeast / Northeast Development Agency. Data on droughts and floods in the Northeast. Message received byregnaldogouveia@bol.com.br on 23 November 2005.

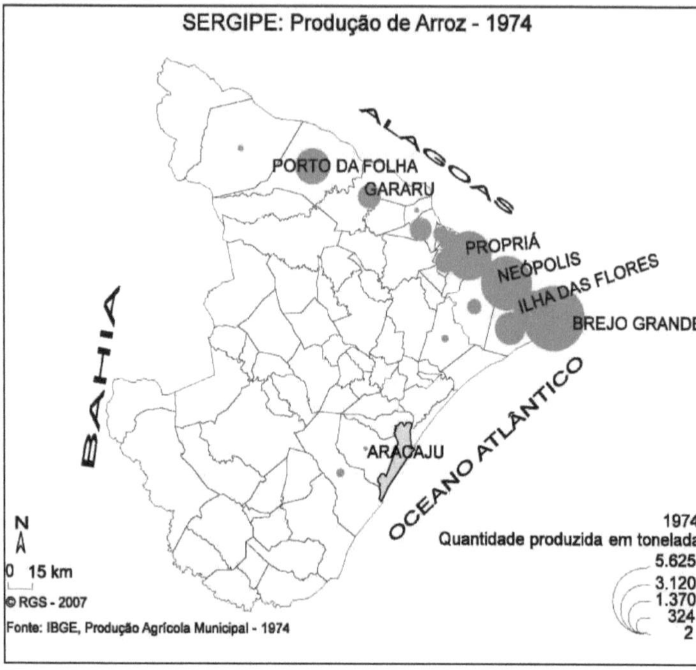

Figura 18. Map of rice production in the state of Sergipe, 1974.

Furthermore, the map (figure 18) corroborates this concept and highlights the municipality's great potential for rice production. According to the map, in 1974 this municipality was Sergipe's fifth largest rice producer and is downstream from the Alagoas towns of Pão de Açúcar and Piranhas. Changing the order of these cities, they are located parallel and respectively to the first two municipalities of Sergipe in West X East, namely: Canindé de São Francisco and Poço Redondo.

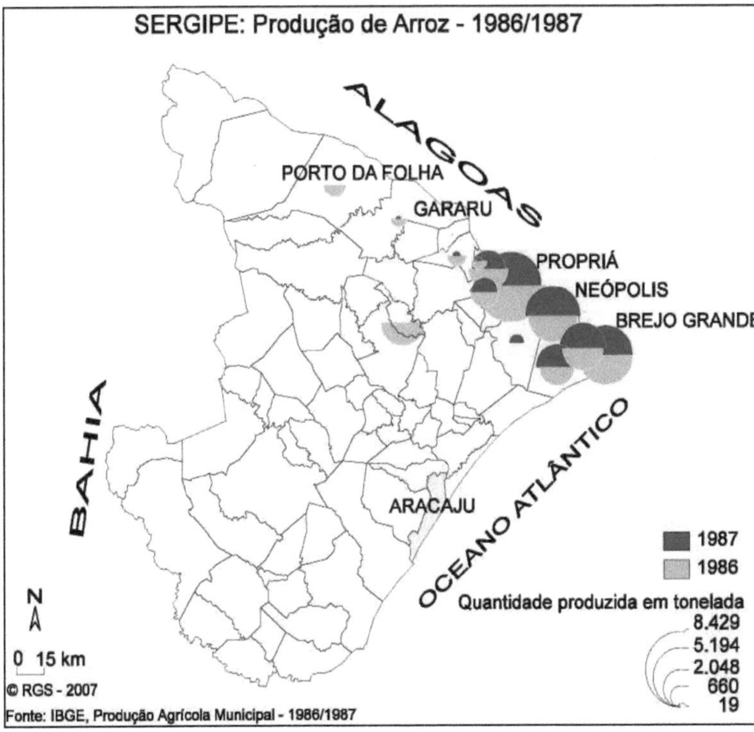

Figura 19. Map of rice production in the state of Sergipe, 1986-1987.

On the other hand, the map (fig. 19) shows an interruption in the rice harvest in 1987. It was from this moment on that signs emerged of the end of the subsistence crop cycle for the Porto-Folhense riverside community. One of the signs of this was the interruption of the natural flood, which meant that the cereal crop was not harvested.

One of the causes, if not the main cause, of this interruption was the start of work on the Xingó Dam, since rice was grown normally before then.

Finally, the extinction was confirmed by reading the map in figure 20.

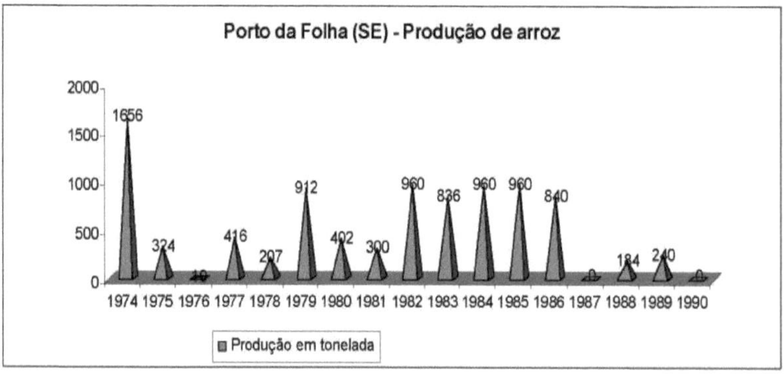

Graph 8. Rice production in Porto da Folha/SE. Source: Ibge, Municipal Agricultural Production - 1974/1990.

Analysing the graph (no. 8) below, and considering that the flood cycle was the determining factor for rice production, we can see that the rice harvest was interrupted in 1987. From that year onwards, production became very irregular, so much so that it showed a very significant drop until its extinction was later ratified. Looking back at previous periods, 1976 records production of just ten tonnes, which means a sharp drop in the rice harvest.

In this regard, it is important to note that "[...] in 1975 it was decided to reverse the previously planned sequence, building the Luiz Gonzaga Plant first and then Xingó" (Enge-RIO, 1993b, p. 05). In other words, this event may also have been the cause of the fall in the rice harvest that year. There was great irregularity in the production shown in the graph, i.e. the figures began to fluctuate sharply after the Xingó HPP was installed. For example: in 1974, 1,656 tonnes of rice were recorded, in contrast to later years, especially from 1987 onwards.

However, if the Capivara River flooded during the period when the floodplain crops were growing, it would be enough to cause a sharp drop in the harvest. This is because the water in the Capivara Creek is brackish, so when the rice crops received this water in the floodplain, the planting would turn yellow and the harvest would be greatly reduced.

It's important to emphasise that, very rarely, this could happen because, often, the voluminous floods of the São Francisco responded to the sporadic floods of the Capivara. It should also be borne in mind that rice cultivation in Porto da Folha/SE did not depend on sophisticated technology. Even so, at the time of the rice-growing cycle, in order to obtain better and greater production, "it was enough to mix a bit of external soil with that from the floodplain and a bit of vegetable bagasse and spread it around" (verbal information)[16] . The interruption of rice production in 1987 was due to the start of civil works on the Xingó HPP. According to Enge-RIO (1993b), these works began in March 1987, but at a slow pace due to financial difficulties, and were resumed in May 1990.

By analysing the map (fig. 20) below, it is possible to see how rice production in the São Francisco valley has changed in the state of Sergipe. The most aggravating aspect of this was the confirmation of the extinction of this production in the municipality of Porto da Folha immediately after the full operation of the Xingó HPP, between 1996/1997. Finally, the last generating unit of the plant, according to Enge-RIO (1993b), was scheduled to come on stream in mid-1996.

The map shows that the last record of the crop was between 1994 and 2003.

According to the maps presented, it was possible to follow the decline to extinction of rice production in the area. It is important to emphasise that these impacts occurred as soon as the Xingó Hydroelectric Dam was completed.

[16] The source has not been revealed for reasons of preservation of rights.

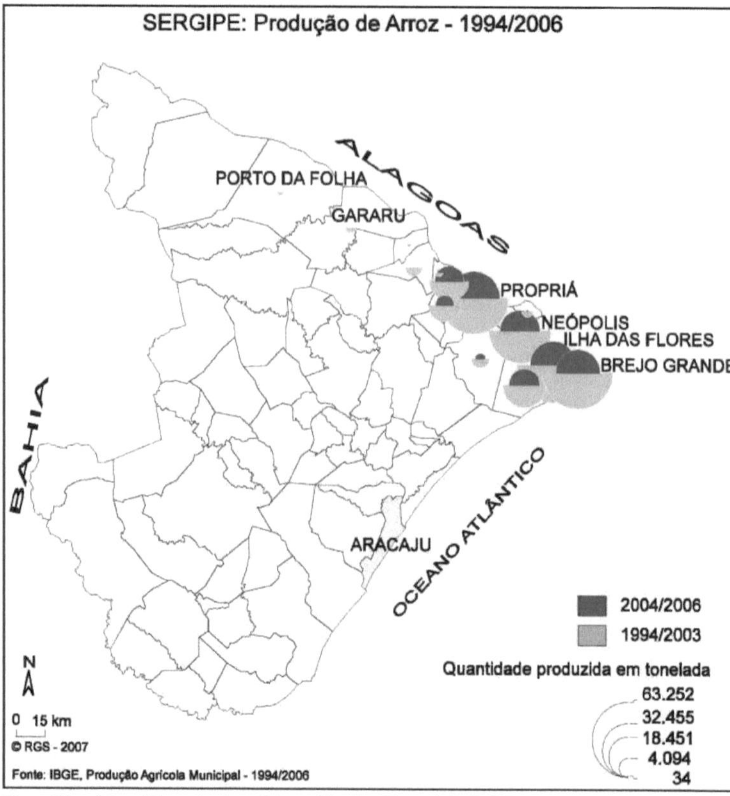

Figura 20. Map of the extinction of rice cultivation, Porto da Folha/SE.

The flood cycle that occurred in the area was based on the existence of the river's natural flow because, due to this flow, the river's natural flow was the same as the river's natural flow.
water level was higher in the course of the year. On the other hand, as already mentioned, with the control of the flow in the dams it was realised that the level of the river was much lower than at the time of the natural flow.

As a result, the social impacts, among other things, were notable in relation to the custom (rice farming) that at one time existed for centuries, passed down from generation to generation. Unexpectedly, these impacts have led to drastic changes in the community's way of surviving, such as inducing them to look for other means of survival. As a result, the river dwellers stopped planting rice in the floodplains and started planting beans, maize, etc. in the rainfed areas when it rains.

It was around "1991, when the floods ended" (verbal information)[17] , that the place suffered serious consequences, such as the worsening famine that hit the riverside community. As a result, everyone "suffered, many people went hungry" (verbal information)[18] . In other words, the social impact was unleashed on local society, among other ways.

The interviewees lamented the end of the "abundance" because it was fundamental to their livelihoods. This abundance is related to the abundance that existed there, from rice cultivation to the cultivation of various vegetables and horticultural crops, provided by the presence of water in the floodplains, which was fundamental for the existence of the immense population of fish that existed in the lagoons and was important for the people's leisure time.

It was possible to observe other impacts after the construction and full activation of the Xingó Dam, with the accumulated water, since "Consequently, this significantly **alters** the physical, chemical and biological processes of the ecosystems present in the region." (SILVA, 2002, p. 29, emphasis added). Since the region encompasses the site studied, it can be seen here that the aforementioned alteration may have been the cause of the drastic decrease in the fish population. This episode was yet another social impact, since it directly affected the local community in terms of access to the protein (fish) that is so necessary for their diet. This assertion is confirmed by Bastos and Fonseca (1997), since fishing was practised as a means of supplementing the riverside community's diet. In other words, fish was served as part of the São Francisco riverside community's meals.

It is possible to say that these impacts would not have occurred in the region studied if the EIA/Rima for the Xingó Hydroelectric Power Plant had taken into account the aforementioned concept, according to Ab'Sáber (2006). As a result, predicting impacts in relation to projects such as the Xingó Dam, among others, as the author rightly observed, requires taking into account the possible future transformations of space and environmental quality, as well as predicting the scenarios and their adversities that may be created in the future. Thus avoiding what happened in the place studied, a transformation of the geographical space due

[17] The source has not been revealed for reasons of preservation of rights.

[18] The source has not been revealed for reasons of preservation of rights.

to anthropic action, incurring a slip by acting on a certain unknown situation.

CHAPTER 4

STATE ACTION AND SOME POLICIES RELEVANT TO THE SOCIO-ENVIRONMENTAL ISSUE

This chapter discusses the actions of the state and its policies on the socio-environmental impacts on the banks of the São Francisco River. To this end, interviews are conducted with some federal organisational representatives and those from the state of Sergipe, where the study area is located. This is followed by an analysis of the policies of the dams installed along the riverbed. At the same time, the EIA/Rima of the Xingó HPP is analysed, highlighting the discussion of the area of influence of projects of this kind and the government's actions to correct the impacts that have devastated the area. In addition, a prospective assessment is made of the regenerative capacity of the local ecosystem, after analysing some of the policies or reactions of state bodies to the area, especially in the Ilha do Ouro floodplain, in Porto da Folha/SE.

THE STATE AND POLITICS

From the point of view of the state and politics, it's important to bear in mind that the term "state" has the meaning of a political-administrative body, which includes sovereignty on a scale ranging from local to regional (state) to federal.

According to Sartori (1997, p. 170-171, emphasis added):

> The democratisation or massification of politics implies not only diffusion and dilution, but also ubiquity. The vertical structure is joined by a horizontal expansion, which once again influences the language of politics. After millennia of relative stasis, how many changes in just over a hundred years! The state is expanding, but political processes can no longer be contained within it: as a result, the concept of the state is broadening, gradually being replaced by the much more elastic concept of the 'political system'.

In this context, the concept of a democratic state has been clarified and it can be managed by representatives promoted by the masses of people. Thus, when the representative of the institutional body takes decisions that are contrary to the mass of the governed, they will soon face a loss of power, such as a drop in popularity, and they may also be highly criticised by demonstrations by the people.

Policies are certain decisions made by people who occupy

political office. As Sartori (1997, p. 172) rightly observed:

> Political decisions refer to a wide variety of subjects: they can be economic, social, religious, educational, legal, etc. If all these decisions are primarily 'political', this is because they are made by people in political positions. This is their political 'nature'.

According to Weber (1982, p. 98), "Hence 'politics' for us means participation in power or the struggle to influence the distribution of power, whether between states or between groups within a state".

If there is peaceful dialogue between society and representatives of the spheres of state power, we hope to find ways of bringing significant benefits to democracy, resulting in gains for both parties.

"Society is not just a distinct 'social system', independent and self-sufficient in relation to the 'political system'. It is more: it is the social system that generates the political system." (SARTORI, 1997, p. 167). In this way, the forecast of impacts can be ratified, because according to Ab'Sáber (2006), this expression takes into account the total space, which includes the social space and its relationship with nature. In this sense, it's important for society to enjoy the right to demand that the state, within its remit, comply with environmental policy laws, in order to preserve the environment in the short, medium and long term, with a view to ensuring a good quality of life for future generations.

It is possible to believe that, through democracy, the lower classes (those currently earning up to 2 minimum wages) and the lower classes will have just as much power in the state as the middle classes and the upper classes. That's if the first two mentioned above don't have the most power, since the contemporary regime of government, democracy, which manages the current Brazilian state, among other countries, depends on the people, quantitatively. From the lower classes to the lowest, the people who make them up represent the greatest number on the face of the earth in the face of the system that governs the economy. Thus, by using the power of information in a broad sense, especially in terms of education, the people, in the form of the law, will be able to make progress in negotiations that are relevant to socio-environmental issues.

A POLITICAL APPROACH TO HYDROELECTRIC DAMS ON THE SÃO FRANCISCO RIVERBED FROM SOBRADINHO TO XINGÓ

All the hydroelectric plants, such as Luiz Gonzaga (formerly Itaparica), the Paulo Afonso Complex I, II, III, IV and Moxotó, as well as Xingó, are managed by Chesf. The main aim of these hydroelectric plants' policies was to guarantee the energy supply of the Northeast region and foster its socio-economic development, thus contributing to the country's development. Meanwhile, there was a plan to build the Sobradinho Dam, so that its purpose was the multi-year regularisation of the São Francisco River. This policy was confirmed in 1979, and the dam is now generating electricity.

As for Xingó's environmental policy, according to Enge-RIO (1993f), its environmental prognosis followed agreements between Chesf, the state environmental agencies of Alagoas and Sergipe, as well as the federal government. This prognosis refers to the identification, description and assessment of the environmental impacts resulting from the implementation and operation of the Xingó HPP. The respective bodies are: Alagoas Environment Institute (IMA/AL), Sergipe State Environment Administration (Adema/SE) and the Brazilian Institute for the Environment and Renewable Natural Resources (Ibama).

The local public policy of the Xingó Dam has maintained the socio-economic development observed with some effectiveness in only five municipalities in the area of influence, Piranhas/AL, Canindé de São Francisco/SE, Olho D'água do Casado/AL, Delmiro Gouveia/AL and Poço Redondo/SE.

The dam's regional policy, in partnership with the governments of the two states, promoted some significant advances, such as investments in the road and waterway systems. Together with the federal government, Chesf has increased development in the area with investments in both the federal road system and the aviation system with the construction of an airport in Paulo Afonso/BA, linking this municipality to important Brazilian cities in the northeast.

It was noted that some environmental plans and programmes were taken on by Chesf, such as the Conservation Unit adjacent to the Xingó Reservoir, observed by Enge-RIO (1993g). However, it can be seen that this environmental policy has not been extended downstream of the dam to the mouth of the São Francisco river basin. Taking into account the negative impacts that the entire riverbank suffered immediately after the dam wall, it can be seen that it would have been necessary to apply the environmental programmes to the areas affected by the impacts.

On the other hand, it's important to emphasise that the local and regional policies mentioned in relation to the Xingó hydroelectric project were limited to the five municipalities closest to the reservoir's surroundings, which, according to the EIA/Rima, is the area of influence. On the one hand, very few projects in the agricultural sector have been developed downstream in the Lower São Francisco, as discussed below. On the other hand, this has left something to be desired because the vast majority of the municipalities bordering this region on the Sergipe side are on the margins of policies to mitigate the impacts of this project. Mitigating policies, in the sense that they could at least come to soften the impacts, in the hope of correcting the damage caused, at least the reversible ones such as the possible regeneration of the São Francisco River's riparian forest.

It's important to note that this framework of ideas emphasises the developmentalist view of the implementation of these projects. In the case of Xingó, it can be seen that some negative impacts were not foreseen in the EIA/Rima, especially with regard to the project's area of influence. Among the various impacts considered by Enge-RIO (1993f) to be relevant, the loss of vegetation cover during the operation phase (power generation) of the plant downstream was not taken into account. This impact had to be taken into account in the same way that the EIA/Rima considered the impairment of water uses downstream of the dam due to the operation phase of the Xingó Dam.

Another significant impact not foreseen by the EIA/Rima was the reduction in regional employment and average family income, when considering downstream from the river to its mouth, during the Xingó operation phase. It was from this phase of the dam that the Porto-Folhense riverside community stopped growing rice in the floodplains because of the regulation of the river's flow. This resulted in a drastic reduction in family income, so that people went hungry.

As for the various impacts expected from the Xingó HPP, according to Enge-RIO (1993f, p. 19) "[...] no significant impacts are expected in relation to erosion of the downstream banks during the reservoir's operating phase". On the other hand, it can be seen that the river's natural flood cycle has always collaborated with the existence of the riparian forest; since its extinction, this vegetation has undergone significant changes. As the riparian forest is fundamental in combating river erosion, the operation of the Xingó HPP, based on its location, has contributed to the advance of this process along the banks of the São Francisco downstream.

Enge-RIO (1993f) states in the Xingó EIA/Rima that the change in the navigation conditions of the São Francisco River is a positive impact, as it will provide 60 kilometres of navigability. However, it is important to emphasise that because these 60 km are limited to the Xingó water mirror, the 179 km stretch from Piranhas/AL to the mouth of the river is no longer potentially navigable by ships or barges. As a result, there was a significant negative impact according to existing records on the contribution of these means of water transport to the development of the municipality in Alagoas, where the museum that holds memorable visual documents is located.

THE UHE-XINGÓ EIA/RIMA AND THE SET OF GOVERNMENT ACTIONS

The set of actions, whether on the part of the state or federal government, or Chesf (the developer of the HPP in question), implemented in the area where the HPP-Xingó project was installed, did not have the desired scope. This can be seen from science and a well-prepared EIA/Rima in accordance with environmental policy law, among other circumstances that will be discussed below. For this reason, it is important to raise the issue of the area of influence considered by this project.

The scientific contribution, in terms of the scope of the surroundings of the project installed, is theoretically evident in this study. In view of this, according to Ab'Sáber (2006), a correct forecast of impacts takes into account the total space.

This is a space that encompasses both nature and society, which man needs to relate to because it plays a fundamental role in organising him over time. Thus, it can be seen that it is difficult for anthropic action to organise space if man himself does not preserve the total space that is so necessary for socio-environmental development.

As Milaré (2006, p. 51) rightly observed:

> It is well known that any development project interferes with the environment, and since growth is an imperative, it is necessary to discuss the instruments and mechanisms that reconcile them, minimising negative ecological impacts and, consequently, economic and social costs.

In spite of everything, the concept adopted by the EIA/Rima of the Xingó HPP to define the area of influence of its project is in line with the study's concept, as follows.

For Enge-RIO (1993d, p. 01):

> The area of influence of a project is generally defined as the geographical space potentially affected, directly or indirectly, by the actions to be carried out during both the installation (construction) and operational phases of the project.

The Xingó EIA/Rima took as a reference the micro-region determined by Ibge in 1989, which includes the municipality of Porto da Folha/SE, to delimit the area of influence of the hydroelectric project. However, Enge-RIO (1993e) did not consider the municipality in question to be a direct area of influence, nor an indirect one.

In this sense, the whole of the Lower São Francisco in Sergipe should be considered as an area of indirect influence of the project, because it is located on the bank of the river affected by significant impacts, but this region was not taken into account. From the Xingó Dam, downstream to the mouth, the river's flow has changed significantly, so much so that after the existence of this dam the highest flood peaks never reached 3,500 m^3/s, whereas before this development the river's flow reached 13,000 m3/s.

The policies here will be analysed in relation to the area vulnerable to impacts, which is the subject of this study, and not restricted to what was considered by the EIA/Rima of the Xingó HPP. As a result of the existence of the hydroelectric project, a number of government plans and programmes have been implemented in its area of coverage, namely: the Hidroagrícola Califórnia project, Jacaré - Curituba, Aproveitamento Hidroagrícola das Áreas Semiáridas dos

Estados de Alagoas e Sergipe, among others.

All of these projects are geared towards agriculture. The first, with an area of 4,000 hectares (ha), was the one that showed significant operational progress, as it increased both agricultural production and the number of families employed in the sector. However, there were no major advances in the other two projects. In contrast to the success of the first, the latter did not have all of its planned area exploited. The area earmarked for exploitation in Sergipe was 95,000 ha and in Alagoas 100,000 ha, but of these areas, only around 4,100 ha and 20,000 ha, respectively, were being exploited by the two states.

The Jacaré - Curituba project fell under the responsibility of the Sergipe government with approximately 4,000 ha of land to exploit. According to Enge-RIO (1993b), the 4,100 ha referred to as the proceeds of the third project (Aproveitamento Hidroagrícola das Áreas Semiáridas dos Estados de Alagoas e Sergipe) correspond to Jacaré - Curituba. This means that the area exploited in the third project is the same as that used for the second, so there is no addition of the area earmarked for development, but a substitution, since they are different projects. In this way, the Semi-Arid Areas Hydro-Agriculture Project did not make significant progress in terms of development, as only a small area of just 100 ha out of 95,000 ha was exploited in Sergipe.

For Enge-RIO (1993b, p. 109):

> When the exploitation of hydraulic potential affects more than one state or municipality, the distribution of the percentages [financial compensation] referred to in this law will be made proportionally, taking into account the areas flooded and other parameters of regional or local public interest (Law No. 7 990/89, Article 5).

As for the parameters of regional public interest, it was noted that in the case of Porto da Folha/SE, financial compensation compatible with the damage caused is possible due to the negative impacts that the municipality has suffered as a result of the existence of the enterprise in question, as already stated.

The biggest and best investments in regional infrastructure policies were notable in the area of influence considered by the EIA/Rima, with improvements to the road system, state and federal motorways, as well as waterways, railways and airports.

According to Enge-RIO (1993e), these roads cross the municipalities in the area of influence. The waterway, on the other hand, is restricted to the Xingó Dam lake, linking that area to the territory of Paulo Afonso/BA. The aviation system has only one airport in Paulo Afonso/BA, which can connect the area with some of the northeastern capitals such as Salvador, Recife, among others, according to Enge-RIO (1993e).

These actions took place jointly between the local, regional and national governments, but with greater financial support from the latter two.

Some of the environmental policies implemented by the hydroelectric concessionaire were noted in the Basic Environmental Project (PBA) developed by Enge-RIO. This project comprises two categories of ecoregion, defined as I and II, and the criteria adopted to characterise II are noteworthy. In this regard, "Ecoregion II comprises the segment of the São Francisco River situated between the Xingó HPP dam and **the** estuary, **and the tributaries of the stretch.**" (Enge-RIO, 1994b, p. 30, emphasis added).

Although the project's geographic scope includes the area in which the Capivara River is located, as it is downstream from the dam, it seems to be detrimental to the aquatic fauna management and conservation programme. Given that the project's criteria are not being met, according to Enge-RIO (1994b, p. 30):

> The general objective of aquatic fauna conservation, according to the criteria adopted in drawing up this project, is to promote the preservation, maintenance and sustained use of the animals that make up the region's aquatic fauna. To this end, in conjunction with the other programmes included in the Basic Environmental Project, this project has the following specific objectives: [...]
> - Enable the perennial utilisation of aquatic species;

It should be emphasised that the Capivara River was observed in the Xingó EIA/Rima as perennial, but today it is common to notice the intermittence of its bed at the height of the extensive river plain in the area. As a result, an aggravating factor has been identified: the Capivara's fish population has literally been wiped out, if we compare to the abundance that existed before the development in question was realised.

In this case, the state needs to take more agile action to monitor these relevant socio-environmental issues.

The agility of the state could also prevent its institutional absence, as happened in the geo-historical case of the relationship between sharecroppers and wealthy landowners in the region. In this sense, Monteiro (1962) noted that in the Lower São Francisco, there was a case of a landowner charging a sharecropper

33.3% interest on an advance, while the sharecropper was financed by Banco do Brasil at 6% per year.

Another aggravating factor was that the project only authorised actions on aquatic fauna. In other words, it didn't consider the whole area, as if the riverside communities only survived on fish.

It's true that the people also fed themselves by fishing, but much more than that, they survived by growing rice. Nowadays, they don't have either of the two activities that used to provide them with plenty.

As a result, according to Enge-RIO (1994b, p. 20, emphasis added):

> The stretch can typically be classified as a lowland ecological zone, particularly after the town of Pão de Açúcar. The lowland zone of a river is characterised by its small slope, low current, sandy bottom and generally muddy water. In this zone the river deposits its sediments and has extensive floodplains on its banks, made up of alluvial and/or hydromorphic soils that are flat and rich in organic matter; marginal lagoons and a vegetation cover made up of riparian floodplain forests and hygrophilous fields. The São Francisco floodplain in [ecoregion II] has many of these characteristics.

Of course, Porto da Folha/SE is downstream from Pão de Açúcar/AL and has river plains, so it is imperative to implement mitigation policies after the hydroelectric project is completed due to the impacts caused.

The archaeological salvage programme began in mid-1988 and lasted until the reservoir was filled, according to Enge-RIO (1993b). It should be borne in mind that this period was unsatisfactory given the site's great relevance to human history. This history is significantly distant compared to the short period during which the Xingo Dam was filled, which took place in the first half of 1994, resulting in a short period of archaeological salvage. The site where these works were carried out could be better explored by science for the benefit of society, in terms of better clarification of human ancestry before the Xingó reservoir was formed.

For Enge-RIO (1993b, p. 12):

> The main objective of the Xingó Hydroelectric Power Plant is to increase the supply of electricity to the Interconnected System
>
> CHESF/ELETRONORTE, in order to support the expected growth in demand for electricity in the region served by the system, thus hoping that the plant will contribute to the socio-economic development of the Northeast Region.

Thus, the use of the electricity system is believed to be one of the main ways of bringing water to the sertanejos who find themselves far from the edge of the

São Francisco, even though they live in municipal territories bathed by the river. Given the efficiency of this system and precisely because of this, it is noteworthy that, for many years over several decades, several municipalities on the banks of the river were in a state of calamity, especially in the first decade following the installation and full operation of the Xingó Dam. This was due to a lack of water. If this life-sustaining liquid existed in all the homes of the sertanejos scattered throughout the Catingueiro region, this would not have been the case.

It is essential to draw attention here to the lack of water supply for the very population located on the banks of the river, which has had its course interrupted by the project whose objective is the socio-economic development of the Northeast.

If only the municipalities on the banks of the São Francisco River in Sergipe had a good water network, especially in their most water-stressed areas, the situation would be very different. These municipalities are bathed by the river and, therefore, it is the right of all residents to have access to water, so this is not unfair, since these territories are close to the mouth, with the furthest being around 180 kilometres away. The implementation of this public policy is necessary, as long as the process is democratic, does not favour private individuals and there is rationing in the use of the resource, such as taxing quotas.

As these people are faced with the great difficulties of water in the Sertão, they scrape by (in the sense of arranging, tidying up) as best they can to survive. The arrival of water in this region, at least through proper channelling, would be the realisation of a dream for the sertanejos, bringing them closer to a quality of life that is at least ideal.

This meaning draws attention to some of the policies being implemented in Brazil, in particular, in this case, government action on the implementation of the Xingó Hydroelectric Dam. For example: its EIA/Rima spelled out the main objective, the increase in the supply of electricity, which it hoped would contribute to the socio-economic development of the Northeast. As we can see, this development has taken place, but for some of the higher social classes in relation to the sertanejos who live scattered in the caatinga, subject to long periods of drought. These people, although they live in the countryside, participate in

Brazilian democracy, at least to elect politicians who represent the nation.

Extensive areas of land such as Várzea Ilha do Ouro, in Porto da Folha/SE, after the full operation of the Xingó HPP, will never again fulfil their productive potential as they did before the project was implemented. In contrast to the hope of contributing to socio-economic development, as mentioned above, the rice-growing cycle, which had always existed prior to the Xingó hydroelectric project, was extinguished because it depended on nature, the natural river.

With regard to the year in which work began on the Xingó hydroelectric project, which preceded the 1988 Constitution, we can see that it is possible for the state to act on the lack of impact mitigation policies in the case of Porto da Folha/SE. Under the country's political regime, this institutional body has the power to make decisions in this regard. Since the government belongs to the people, it is possible that it fulfils the public needs (public policies) of the population itself. So, given that the EIA/Rima evaded contemporary environmental reality due to the lack of certain laws or revisions to those ratified in 1988, the state can act on the aforementioned case with socio-environmental policies aimed at the good of the people. The state can also correct the negative social and environmental impacts that the site has suffered through the application of these actions and, in this way, it will be possible to confirm its presence there through its governmental institutions.

It is imperative to correct the environmental impacts, because in this way it is possible to reverse the social impacts. By recovering plant biomass, it will be possible to increase the relative humidity of the air through evapotranspiration and, consequently, regenerate the local ecosystem. However, if nothing is done about these policies, the environmental and social impacts tend to persist in the area, preventing full progress in social, economic and environmental development. In this way, the necessary and urgent regeneration of the ecosystem is unlikely to take place. After all the environmental impact the site has suffered, despite having to exert a lot of effort to reverse the situation, the recovery of this physical and natural space is still possible. For this to happen, there needs to be political commitment from the state to implement effective and efficient actions aimed at the environmental and social sectors. In addition, there needs to be investment in research to monitor and promote the progress of the ecosystem regeneration process, as well as the implementation of socio-environmental education to raise awareness and preserve the environment.

DISCUSSING SOME POLICIES RELEVANT TO THE SOCIO-ENVIRONMENTAL ISSUE AND THE ROLE OF THE STATE

It is well known that the main function of the Xingó hydroelectric project is to generate hydroelectric power with a potential of more than 10 Megawatts (MW).

It is known that, according to a reliable source19 , the licensing process for the project in question began in 1993/1994. The first licences were issued as follows: the Preliminary Licence (LP) was issued in 1994, the Installation Licence (LI) in 1997, and the Operating Licence (LO) in 1999.

According to Milaré (2006, p. 51):

> Among the instruments for making development and environmental protection compatible, special attention should be paid to the Preliminary Environmental Impact Study (EIA), which must be drawn up before installing any work or activity that could potentially cause significant environmental degradation. This important environmental planning and control instrument stems from the principle of taking the environment into account when making decisions and advocates the basic obligation to take the environmental factor into account in any action or decision - public or private - that could have any negative effect on it.

According to the author, the LI must be granted after the EIA/Rima, i.e. it must be drawn up and approved before the licence is granted. However, this did not happen in real time with regard to the construction of the Xingó HPP, since the second document was drawn up in parallel with the installation of the project.

Although, for Enge/RIO (1993b, p. 107, emphasis added):

> The Xingó Hydroelectric Plant Feasibility Study, which began in July 1981, was completed in 1982, prior to CONAMA Resolutions 001/86 and 006/87, and all the requirements of the licensing legislation in force at the time were met. It can therefore be concluded that the work in question is exempt from the
> Preliminary Licence (LP) and Installation Licence (LI), having passed the application stage.

The project in question began in 1987, after Conama Resolution 01/86 came into force, but at no point does its text refer to any feasibility study. After analysing the Resolution, it can be seen that in order to carry out works of the size of the Xingó hydroelectric project, as of 23/01/1986, its EIA/Rima must first be

19 The source has not been revealed for reasons of preservation of rights.

submitted for approval by the competent environmental agency. Given this, it can be seen that, at the time, it was up to the state to enforce the requirements of Conama Resolution 01/86.

What was considered here by the EIA/Rima of the Xingó HPP is unreasonable when taking into account the law in force at the time, since according to Conama Resolution 01/86 (1986, emphasis added):

> Article 2 - The licensing of activities that modify the environment, such as the following, will depend on the preparation of an environmental impact study and respective environmental impact report (RIMA), to be submitted for approval to the competent state body and IBAMA on a supplementary basis:
>
> [...]
>
> VII - Hydraulic works for the exploitation of water resources, such as: dams for hydroelectric purposes, above 10MW, sanitation or irrigation, opening of canals for navigation, drainage and irrigation, straightening of watercourses, opening of bars and inlets, transposition of basins, dykes;

In other words, it can be seen that, in accordance with Conama Resolution 01/86, the licence for the project in question could only be granted after the EIA/Rima had been drawn up and approved, as the work began in 1987, a date after that Resolution. Furthermore, none of the volumes of the EIA/Environmental Report date from before 1993, so much so that the first Preliminary Licence was issued in 1994.

Thus, it was impossible for the Xingó hydroelectric project to have been exempted from the LP and LI because its EIA/Rima had not been approved, as it dates from 1993/1994, according to Enge-RIO (1993a). It becomes

It is inconvincing to refer to a feasibility study that began in 1981, given that the archaeological salvage work lasted such a short time. It is therefore possible to believe that this activity, which took around six years, needed more time to be properly exploited than was actually realised.

The fact that the LO was issued in 1999 is an aggravating factor, since this licence was issued after the date of the dam's closure, so much so that in that year the Xingó HPP was already in operation and had therefore completely finished its construction. Taking into account the period in which the reservoir was formed, in 1994, the LO for this project contravenes Article 4 of Conama Resolution 06/87 (1987, emphasis added), since:

> Art. 4 In the case of hydroelectric projects, respecting the peculiarities of each case, the Preliminary Licence (LP) must be requested at the beginning of the feasibility study for the Plant; the Installation Licence (LI) must be obtained before the Bidding

Process for the construction of the project and the Operating Licence (LO) must be obtained before the dam is closed.

With regard to conducting emergency policies, there is a certain complexity, and it is not as simple a manoeuvre as the source consulted made clear: in a meeting with other institutional public bodies, I learned that they warn by radio and television when the flow from the dam increases downstream. It is well known that all the riverbank dwellers are vulnerable to the contemporary flooding of the river, whose flow is now artificialised. However, they never know when the floods will occur, since the natural cycle no longer exists and, as has been said, not everyone has a communication device to keep them informed. What's more, there are many riverbank dwellers buried in the caatinga on the banks of the river, points that are practically impossible to reach by radio waves and TV transmissions. As long as the natural flood cycle existed, they knew when it occurred by force of nature, because there was a natural predetermination of the phenomenon on an annual basis, unlike today.

All of this corroborates the fact that the state, through its institutions, was aware that controlling the flow of the São Francisco River, especially on the part of the Xingó HPP, would cause some impacts on riverside communities. According to this source, there is a study with questions to this effect in the Public Prosecutor's Office.

It is also known that there is a commitment, an environmental compensation, i.e. possibly the creation of a Conservation Unit. This would cover the area from Canindé de São Francisco/SE to Porto da Folha/SE, but it has not materialised.

According to this source, a certain organisation (identity withheld) contemplated, in the EIA, the control of the flow of the São Francisco River by the Xingó Dam, in such a way that it would promote artificial flooding without causing significant impacts on society and that this would occur almost naturally. According to the information, there is certainly a lawsuit against this body, since it has not honoured the consequences of the flow in the riverside area. This is why the unidentified source or its representative is pursuing the case with the Public Prosecutor's Office. In spite of everything, in December 2007, a certain licence had not yet been renewed, but its renewal would take place. It was noted that this licence is, of course, the LO for the HPP project in question.

In a Monitoring Report from 2007, from the Department of Environmental Assessment and Monitoring (Deamoam), the source of which has been preserved, it is clear that the state is aware of the problems (impacts) affecting the riverside population due to the control of the flow of the São Francisco River at the Xingó Dam. Likewise, in this report, it is unequivocal that a certain public body in the state of Sergipe is aware of the responsibility that another certain body has in relation to mitigating social impacts on riverside communities. Despite this, there was no evidence of any kind of policy being implemented in this area.

The EIA/Rima in question was drawn up simultaneously with the construction of the Xingó HPP, as this study found that the LP was issued in 1997 and the LI was already in place in 1994.

As noted by Milaré (1993), the LI must proceed with the preparation and approval of the EIA/Rima. In the same way, Conama Resolution 001/86 considers the procedure for this licence, since, according to this resolution, it can be seen that the licence refers to the process of all the licences for the project, but not just one or two licences, as mentioned by Enge-RIO (1993b). In this regard, a summary of the licensing process for projects of this nature is presented, in the form of the Law (see ANNEX E).

An important governmental action that needs to exist efficiently and be demanded of Chesf by public bodies and the people is the existence of an emergency policy, but one that is effective at least during the alert period of the Xingó Dam. That this policy be applied to the entire lower course of the São Francisco River. It is possible to believe that this would avoid irreversible impacts such as the death of people from electric shock, in this case from contact with the water, due to the current unexpected floods, as complained about by the local population.

However, neither this policy nor others of great importance have been applied in the region in a way that clearly considers the total space, as Ab'Sáber (2006) rightly observed, at least in relation to all the riverside municipalities that are vulnerable to these impacts. This total space encompasses both the social environment and nature, which man needs to carry out his activities.

It is known that the Integrated Protection Unit has not yet been set up. This would be one of the necessary environmental policies aimed at correcting the impacts caused by the Xingó HPP project, but it has not been implemented.

With regard to the renewal of the LO and taking into account the knowledge of the intense impacts caused by the dam on the riverside communities of the Lower São Francisco, everything indicates that there was a possible lack of institutional power on the part of the state. From this perspective, Teixeira et al. (2006) rightly observed that states and municipalities are losing institutional power. This cannot continue to happen because the contemporary state is governed by the democratic system and, in order to corroborate its fullness, as has happened and continues to advance in this direction, it acts on behalf of the people.

In this way, people can demonstrate in accordance with the law and demand their rights, especially the right to live in harmony with the environment, both in social and environmental terms, in a fair and respectful way.

In this regard, according to Théry and Mello (2005, p. 80), the "[...] council [Conama] is the centrepiece of the National Environmental System (Sisnama). In order to set up this system, environmental councils must be set up in all Brazilian states and municipalities". According to the authors, the role of the Special Secretariat for the Environment, with its institutionalisation from 1973 until it existed around the 1990s, was to establish a national environmental policy by integrating local bodies into this national institution. That said, there is a way of looking for solutions to problems such as the one studied, starting from the total space towards the local. In other words, it is possible that an appropriate environmental policy could come from the Ministry of the Environment and, through municipal environmental councils, be applied to municipalities.

According to a second reliable source,20 the action of a certain public body in terms of representing the federal state was to act in the place with the implementation of settlement projects and that, in this way, it would be promoting a democratisation of access to land.

It was possible to observe that the work of this organisation in Porto da Folha/SE began between 1986 and 1987, when work began on the Xingó hydroelectric project.

20The source has not been revealed for reasons of preservation of rights.

According to this source, the Ilha do Ouro farm was expropriated in 1987 due to popular mobilisation with the active participation of the church, which was also the cause of the agency's intervention with its political action in the region.

In addition, from that moment on the families settled there took part in a government programme through public policies, but to this day the agency's representation still encounters difficulties in implementing these policies. However, he compared the situation of the families before (this situation must inevitably refer to the cyclical period of rice cultivation) and afterwards, saying that there had been a substantial and considerable improvement in this regard. He mentioned that the families now have access to credit, the opportunity to work and technical assistance, but the irrigation project that had begun to be implemented in the area had presented difficulties.

It can be seen that the state, represented by the aforementioned body, is aware of the socio-spatial transformation that has taken place in Sergipe's riverside municipalities along the lower course of the São Francisco River as a result of the project in question. It also knows that the floodplains in this region underwent a process of change with the creation of the Xingó Dam. This change limited the number of people who could work in the floodplains of the Fazenda because it would be insufficient space for many people to make a living if everyone who survived on agricultural activities in the area were to be settled.

A relevant proposal for political action by the state government in partnership with the federal government was mentioned, that is, if it is actually implemented. This is the construction of the Xingó Canal, which will bring water by gravity from Paulo Afonso/BA to supply municipalities in the Sertão region, for both human and animal watering, as well as enabling irrigation in areas along the canal. It has also been suggested that there is another irrigation project covering the Sergipe municipalities of Canindé de São Francisco, Poço Redondo and Porto da Folha, which will use water from the Xingó Lake (Dam). Finally, it is believed that this is linked to the transposition of the São Francisco River.

This action aims to develop activities throughout the year, so that there is a link between farmers, with certain crops, and livestock farmers, with the use of crop residues in animal feed. Crops such as rice and maize are irrigated so that the residual straw can be used as cattle feed, thus developing dryland activities as well as farming.

According to this source, the following is not believed to be the best form: the question of the sharecropping, in the sense of the crop being divided into two parts, one for the owner of the land and the other for the ploughman.

When he refers to the landowner's domination over the farmers, who would not be free to vote for whomever they wanted, this would not be a question of the share of the land as such.

There is no doubt that this discussion is relevant. Now, as the state settles families in certain places, it is expected to actually promote the development of the social function of the land. In other words, just having land with no infrastructure will not bring about the development expected and deserved by the people who depend on the land to grow crops and survive on it.

It was noted that one of the biggest historical problems in the region was the lack of implementation of a labour tax policy in relation to the exploitation of peasants. The sharecropper dependent on rice cultivation in the Lower São Francisco was vulnerable to this situation, because as Monteiro (1962, p. 76) rightly observed:

> Despite being perfectly aware of the dispossession of which they are victims, rural labourers are forced to submit to him [the wealthy landowner] in order to obtain new sharecropping contracts. If, for example, the sharecropper refuses to sell his rice to the landowner, he will lose any chance of new contracts in subsequent years. He will be considered a bad person and no landowner will want to employ him. All he can do is emigrate.

This subjection occurred either in the activity of these sharecroppers, who cultivated crops on other people's land, or in the conditions observed by the author. Despite this, the owners were subsidised by the state institution at very low interest rates compared to those they charged the sharecroppers, because according to Andrade (2004, p. 143):

> A threshing machine hulls around 40 alqueires [see alqueire from Sergipe in the second chapter] a day. It's worth noting that it is sold by the São Francisco Valley Commission, now CODEVASF, in instalments, with a down payment of 25% and the rest paid over three years at 7% interest per year. As you can see, the purchase of the threshing machine is an excellent use of capital for the owners.

With regard to environmental protection in the area, the second source made it clear that it has established reserve areas such as the implementation of

Conservation Units. In addition, it has established the recovery of degraded areas such as Permanent Preservation Areas, riparian forests and even has resources for their recovery in the state of Sergipe.

It is sceptical to believe in the democratisation mentioned above by this source when we don't see projects developed through the application of efficient agricultural policies in settlement areas. This situation was observed in the case of Porto da Folha/SE, especially in the so-called Ilha do Ouro and Comunidade Emburaninhas settlements. In this respect, their location may justify this disbelief, since the former is just a few steps from the riverbed and the latter a few metres away. What's more, the latter settlement, which is further away from the river than the former, has not even installed the necessary hydraulic network for animal and human watering and for essential agricultural activities. As for the first settlement, despite being the oldest, it doesn't even have an adequate irrigation project to mitigate even the slightest of the immeasurable impacts that have hit the area.

With regard to the previously mentioned difficulties in implementing public policies, there are currently no solutions. As a result, the floodplains, which were extremely important for rice production and where it was possible to grow the cereal whether it rained or not, are now politically helpless.

Although this source previously reported that the Ilha do Ouro Settlement has technical assistance and substantial improvements, it was noted that these were not confirmed. The reason for this is that no project was observed in development there. Because of the situation in which the Ilha do Ouro floodplain finds itself, it is not believed that the state has applied the necessary policies there to correct the impacts that have affected the socio-environmental environment.

Considering the number of people who benefited and those who depended on the activities carried out in the Ilha do Ouro floodplain, not all of them were settled because the contingent was so large. As a result, we can see the inefficiency of the state policies that were put in place, given that the Várzea had space for one hundred and four families to produce. Lately, however, only 94 families make up the settlement. Despite the smaller number of people possibly working, there has been no success in the area in terms of the situation previously pointed out by this source.

If the state really wanted to achieve social justice in the area, it would have to at least expropriate all the land from the bank of the Capivara to the municipal centre, since this river or stream is susceptible to the floods of the São Francisco. Consequently, it would have to apply the necessary policies, but with efficiency, responsibility and respect for the socio-environment in an effective way.

Efficiency can be seen, among other things, when, according to Sene and Moreira (2000, p. 284):

> The struggle for agrarian reform doesn't end with the conquest of land by the workers or the realisation of settlement projects by the government. There are other fences that need to be torn down. These are the fences that prevent the existence of an agricultural policy geared towards the interests of family farming.

Policies of responsibility and respect, on the other hand, can be promoted in the fulfilment of the role of the state, in the sense of socio-economic justice and environmental preservation or recovery of the environment when degraded. There are situations in which the public institution has the resources to implement certain policies, but for some reason fails to do so.

The construction of the Xingó Canal could be a relevant state action from the point of view of this text, in terms of mitigation policies, as long as it is carried out in compliance with the parameters of national environmental policy in the form of the law. In this case, attention is drawn to the mention of the transposition of the São Francisco River. In other words, it is only possible to believe that the Xingó Canal will be realised once this state action has actually been implemented. The reason for this disbelief is that, when the Xingó Dam was installed, according to the first undisclosed source, a certain body was to comply with a certain flow of the river from the dam, but this has not been done.

In this context, and if the project is realised, it can already be mentioned that due to the lack of concrete effective policies, there has been a possible historical debt owed by the state to the place. It has been observed that there is no effective agricultural or environmental policy in place, with government action.

When the second undisclosed source refers to the possibility of farmers liaising with cattle ranchers, promoted by the aforementioned Canal, it is clear that the state is aware of its competences. However, nothing, if anything, has been done to develop socio-environmental policies. There have been no government actions to promote social development and correct the impacts that have affected the entire area.

The state's inefficiency in implementing decent agricultural policies in the semi-arid region has been realised. The dryland area is in dire need of solutions for its development. Solutions can be found with the arrival of water for the necessary activities, as long as they are carried out competently and with respect for the socio-environmental environment.

It's impressive to see how the state, in a way, has good guidelines for implementing important policies in the semi-arid region, although these are not ratified.

It was noted that there are resources available for the implementation of certain environmental policies, as previously stated. On the other hand, it is noticeable that they are not applied, showing that there is certainly a lack of commitment on the part of the state in relation to the applicability of such policies. When there is political interest in a particular action, such as the transposition of the São Francisco River, it is carried out quickly and efficiently. Even though this action runs counter to science in the sense that it could henceforth cause immeasurable impacts, in the medium and long term, on future generations and, in the short term, on contemporary ones.

The lack of implementation of the necessary effective environmental policy leads us to believe that the state lacks competence, respect and socio-environmental responsibility towards the local community, from the public institution to the local, regional or national level.

What is striking here is that one of the main mottos of agrarian reform is to fulfil the social function of land use. However, as the various settlements in Sergipe's Alto Sertão have been developing, they do not fulfil this function, as there is no adequate infrastructure for at least a hydraulic network, which is extremely necessary for agricultural development. A stone's throw from the River São Francisco is the Ilha do Ouro settlement, which contains part of the extensive floodplain that has contributed to the socio-economic development of Porto da Folha/SE. After having gone through this process (settlement), during the dry season, the Ilha do Ouro floodplain looks like a wasteland due to the lack of infrastructure, for example. So it's hard to believe that the state is fulfilling its role of promoting the fulfilment of the social function of the land for small landowners. If this function is confirmed, the institutional body, through its political representatives, can ratify the dream of economic democratisation through its action as a regulatory agent, a role that also falls to it.

In other words, water for agricultural activity does exist, but there is no discernible policy for managing this natural resource. In this case, turning to the sustainable management of reservoirs, Silva (2002, p. 46-47) rightly observed:

> [...] unmeasurable costs and benefits must be the subject of social and inter-institutional negotiation, in which interests related to citizenship, public health and safety, environmental protection and economic and social development must prevail [...]
> In this sense, public authorities play an important, if not central, role in the negotiation process mentioned above. It must endeavour to regulate this process and make the benefits and negative impacts transparent to all stakeholders.

According to a third reliable source21 , the policy of a certain organisation was to develop projects to mitigate impacts in the area, extensively during and after the period in which the Paulo Afonso/BA hydroelectric complex was built.

According to this source, it was clear that a study had been carried out involving seventy-six floodplains in the Lower São Francisco, ranging in size from thirty to forty hectares. Of the seventy-six, four large floodplains were in the state of Sergipe, and these were the ones to benefit from the projects. These floodplains are: Propriá, Cotinguiba / Pindóba, which is actually two but is considered one large floodplain, Betume and Brejo Grande. What is striking here is the fact that the two floodplains represent a single floodplain. This may have happened so that policies could be implemented there, since if weren't joined together, government actions might not be implemented. Thus, it can be seen that, as with the EIA/Rima for the Xingó HPP, the necessary impact mitigation policies were not implemented there by the body behind this discussion.

According to the table (no. 3) below, the set of government measures applied to these four floodplains also applied to Porto da Folha/SE because of its productive potential.

Table 3 - Sergipe: municipalities with the highest rice production in tonnes (T) - 1920

21The source has not been revealed for reasons of preservation of rights.

Municipalities	Quantity of rice produced in (T)
Propriá	1.684,1
Vila Nova	1.333,5
Porto da Folha	625,9

Source: Ministry of Agriculture, Industry and Commerce. Census of Brazil - 1920

The table (no. 3) above corroborates this assertion because of the important rice production in Porto da Folha/SE, which was ranked third in the state in 1920 and, until the day before the Xingó HPP came into existence, accounted for the fifth largest production of the cereal, as previously stated.

As for the precise start of the agency's intervention in Porto da Folha/SE, there was no accuracy, because according to this source, its function was not limited to one location, but that it began working in the region more effectively in 1970.

In order to control the flow of the São Francisco River after the Paulo Afonso Complex hydroelectric project, which took place long before the Xingó HPP, the agency began to intervene in the river's lower basin. According to this third source, the initial idea behind the agency's actions was to protect the floodplains by building dykes. It also opted to expropriate the protected area in order to distribute it to beneficiaries who depended on the land, thus calling the resulting project a colonisation project.

It's important to emphasise that there is a solution to the problem, and it would indeed be ratified if these policies were applied with competence, respect and responsibility. In this regard, it was stated that all the major projects have been realised such as the development of irrigated perimeters, as well as the Várzea de Propriá/SE, but the Ilha do Ouro area is not included in these plans. Furthermore, the Ilha do Ouro area is not a project, although it is within the projects with people from there working in the large floodplains such as Betume, Cotinguiba - Pindóba, or Propriá/SE.

In this way, there was a lack of consideration for the case of Porto da Folha/SE, given that this municipality once accounted for the third largest rice production in the state. This can be seen as a lack of responsibility and respect when it comes to the few policies applied there. Hence the reference to a historical debt owed by the state to the area, as has already been stated.

The policies implemented by the agency were intended to mitigate the negative impacts of the Paulo Afonso hydroelectric project. These actions were well-intentioned, but in terms of their applicability they were not well targeted or effective. In other words, the floodplains of municipalities like Porto da Folha/SE and Gararu/SE were not benefited in the same way by these policies, as were those of Propriá/SE, among others in the Lower São Francisco.

Nor have these policies been applied effectively and efficiently since the installation of the Xingó hydroelectric project. Despite this, this project has been the cause of immeasurable and irreversible negative impacts on the socio-environmental environment of the lower reaches of the São Francisco River.

According to this third source, the agency's intervention in the region was to mitigate the impacts of the state's actions to control the river's flow, which is in line with the intervention mentioned by the second source, who said that it had come in response to mobilisations by the people demanding land reform.

In contrast to what the third source reported earlier, it was noted that none of the partner families in the Ilha do Ouro settlement carry out their activities in any project outside the settlement. On the contrary, the community complains about the lack of efficient and effective policy plans in the settlement.

In addition, the economic conditions of the community, which cultivated rice in order to survive, and the difficulties that the river dwellers encountered in travelling between Porto da Folha/SE and Propriá/SE, due to the distance, corroborate the above assertion, since this municipality was pointed out by this source as one of the destinations for the river dwellers to carry out their activities.

The application of some environmental policies in the municipalities bordering the Lower São Francisco still leaves something to be desired given all the time that has passed in relation to the hydroelectric projects installed on the river, especially the Xingó project. Other necessary environmental policies include the preservation of the main river in the catchment area and the treatment of all residential and industrial sewage before it is discharged into the São Francisco.

According to this source, the state of Sergipe has failed to implement these policies in the past. For more than twenty years, projects to develop small floodplains have been proposed, but nothing has been developed. In this way, it can be seen that the policies applied in the Lower São Francisco are ineffective, especially in the case of Porto da Folha/SE as its main floodplain is currently located, in terms of developing agricultural projects.

It was possible to observe that a certain body, according to information from the last undisclosed source, pointed out a way out of the consequences caused by the Brazilian state in carrying out hydroelectric projects on the São Francisco River. It is therefore up to the state itself to correct the impacts it has caused, with a view to its own development. More than that, the people, who have rights in relation to this political and administrative body, need the impacts to be corrected. For this reason, the proposed solution would be effective if it had been planned so that all the floodplains would continue to be flooded naturally and all the owners of the extensive river plain in the area would be expropriated and compensated in real time and amounts.

However, the policies were not implemented in this way. The policy indicated by both this source and the second, land reform, was neither efficient nor fair, at least in the case of Porto da Folha/SE. Firstly, because there was no planning to continue the natural cyclical flooding of the floodplains, which had immeasurable environmental impacts. Secondly, because the potential of Porto-Folhense's extensive river plain was not considered with the same importance by the particular body, according to the third source, in the application of its mitigation policies. In this respect, the local community has suffered various social impacts. According to this study, several people went hungry and struggled to survive after the Xingó hydroelectric project was built.

Here it was possible to see that the state is aware of its competences, but there are several where its omission is evident, as in the case of Porto da Folha/SE.

Historically, the Porto-Folhense community's development model took place without the need to restrict the use of the waters of the São Francisco. Now, if a particular state action has led to the interruption of a form of community development that is not very offensive to the environment compared to some current activities, the same institutional organisation is expected to bear its responsibilities. These responsibilities are, in the economic and socio-environmental sense, for the place affected, since its interest has been achieved (UHE-Xingó) with the extinction of a community's means of survival.

Finally, on the one hand, it was noted that the Xingó EIA/Rima implemented policies only in the area of influence considered by that document. On the other hand, the São Francisco and Parnaíba Valley Development Company (Codevasf) implemented agricultural development policies in a region far from the place that was hit hard by the impacts of the Xingó Dam.

It is worth emphasising that various questions have arisen throughout this chapter, all of which point to the fact that, in the case of Porto da Folha/SE, the state has failed to act.

It is therefore important to emphasise what Rolston III (2004, p. 325) rightly observed:

> This requires policies and behaviours that keep the population and development in harmony with natural environments. It will be difficult to maintain peace between people until we achieve peace with our environment. What we want is not just 'riches', but a 'rich life', and proper respect for the earth's biodiversity enriches human life.

It is possible for the "rich life" mentioned by the author to continue to exist, as long as there are viable policies that are respectful, responsible and efficient in every sense, such as oversight by institutional leaders.

CHAPTER 5

CONCLUSION

From the geo-history and nature of the place, its natural dynamics and the rice cycle, its socio-environmental impacts, to analysing the action of the state with its policies, it has to be considered that government institutions have not shown any commitment to solving the problems that are dragging on in the region. Despite this, it is notable that the state is aware of the situation and is capable of minimising these impacts.

By analysing the geo-historical process and the action of nature in the region, it was possible to conclude that the spatial occupation of the place was closely linked to the discovery of the São Francisco River and, consequently, the exploration of its banks by the first European navigators. In addition, the Capivara River stood out because its waters flowed into the "Old Chico", which facilitated the clearing of the lands of Porto Folhenses. It was also because of this local geography that Porto da Folha/SE was on the banks of the São Francisco River more than it is today, due to the natural flood cycle.

In this context, it is believed that the place was not named at the time of its political and administrative emancipation, but has retained the name of the extensive lowland region bathed by the São Francisco River since the first explorations of its banks, on the Sergipe side, rescuing toponymic information from old maps.

The distinction between the nature of the place and the entire semi-arid region was characterised by the natural richness of the local flora and fauna. This was limited to the natural flood cycle of the São Francisco River, resulting in a local natural cycle, i.e. the surplus water directly interfered with the existence of the dense tropical hydrophilous floodplain forest, water outcrops, etc. All this contributed to the existence of the fauna that lived there. In short, this natural cycle created the means of survival for the riverside community. How the nature of the place is degraded is striking. This shows that, although it's too late, the authorities in charge need to take action on the case, as well as carrying out regional and local studies increasingly advancing in relation to the socio-environmental issue.

The fluvial action in the local floodplains enriched the soil due to the self-fertilisation process of the San Francisco floods. Among other things, this emphasises the need for the flood cycle to be restored. Who is responsible for this? In this case, we believe it falls to the state, a capable institution.

The techniques used by the riverside sertanejo were not advanced, but adapted to their way of life in contact with the fluvial dynamics of the São Francisco. For this reason, the relationship between man and nature in the area was simple, but skilful in terms of rice cultivation, from preparing the seed for sowing, through development to harvesting and, finally, processing, which began and ended at home. Simple, because it was in their homes that the farmers of Porto-Folhenses soaked the rice in water to sow it in the floodplain the next day. It was also in their homes that the vast majority of them husked the rice in wooden pestles using their manual labour.

The land was prepared in the same way, as the farmers mixed the soil from outside with that from inside the floodplain, which was intensely fertile, in order to obtain better harvests, as well as selecting the best-yielding seed for the same purpose.

It's important to emphasise that the techniques used by the Porto-Folhense riverbank dwellers during the rice cycle were literally harmless to the environment, compared to those used in the contemporary world, such as garden agriculture in Asia. This type of agriculture, prominent in Japan and some of China's coastal provinces, was driven by capitalisation as a result of modernisation. The capitalisation referred to here is directly related to agriculture, since it involves the use of advanced techniques in the field, as well as the use of machinery, pesticides, soil improvers, etc.

The labour relationship in the floodplains of Porto da Folha/SE was marked between sharecroppers and wealthy landowners.

The sharecroppers carried out their activities on private land and had to divide the crop they had grown into two parts, from the preparation of the land to the harvest, in short, they kept one half and the other remained with the landowner for the land use concession.

The wealthy landowners, who owned the land and had their sharecroppers subject to them, cultivated rice in surplus. On the other hand, the vast majority of the population, from sharecroppers to many small landowners, grew rice for subsistence.

The following fact is noteworthy here: the greater the number of young children, the more power the sharecroppers had among themselves and, likewise, the small landowners. This is related to the use of manual labour that was typical in those days. In this sense, the larger family produced more rice and consequently had more to eat throughout the year until the next harvest.

In view of this, the space in which the Porto-Folhense riverside community lives and with which, until the end of the 20th century, it related, developing its way of life, comprises the environment as part of nature.

The socio-environmental impacts that occurred in this sphere affected both physical and social space. It was from the total space that the riverside people developed in the place.

Environmental impacts, the consequences of anthropogenic action on the environment, were triggered by the lack of water in the extensive river plain. This occurred after the natural flood cycle of the São Francisco River was extinguished. The main cause of this episode was man's action on the riverbed when he built several hydroelectric projects along its course. It's important to emphasise that among the various hydroelectric projects from Três Marias to Xingó, this one drew the most attention, because it was when it was built that the São Francisco's flood cycle was extinguished. Because natural floods hit the river plains of Porto da Folha/SE at certain times each year, the natural phenomenon was characterised as cyclical. On the other hand, after the existence of the Xingó HPP, with its control over the river's flows, the floods diminished from year to year and, no longer being periodic, the natural cycle was extinguished.

As a result, various impacts have been unleashed on the local environment, for example: one of these impacts that has attracted the most attention is the reduction of water in the subsoil of the floodplain due to the absence of surplus water. As a result, the vegetation, which was present until the end of the last century (XX), has been suppressed, and the water outcrops no longer exist. Another obvious impact is the increased presence of salt in the soil of the main floodplain, which could make the area unviable for growing winter crops because the summer ones no longer exist.

In addition, there were several other social impacts such as losses in the culture of the local community, which was directly linked to the way of life of the sertanejo ribeirinho. At the time, the local people's relationship with nature expressed their way of life, while the local natural aspects, being distinct from the rest of the semi-arid region, promoted this relationship due to their enclave characterisation. In this sense, you can no longer see the cart drivers dragging the rice from the extensive river plain to the city, nor can you hear the sound of the pestle on Rua de Cima. Much less is it possible to see the people bathing in the waters of the São Francisco, which bordered the urban area during the festive season, Christmas and New Year. In this respect, there are only traces of the customs associated with rice cultivation, and the only memories left for the people are the important figure stamped on the altar of the Mother Church and some of the pestles used at the time to process the cereal.

The fact that the study area is affected by all these impacts corroborates the lack of inclusion in the area of influence of the Xingó hydroelectric project. Although the EIA/Rima considered a small area of the São Francisco river basin to be of direct influence, albeit inappropriate from the point of view of this study, it seems to have sought to fulfil the requirements of the law. Nevertheless, it was also drawn up at the same time as the hydroelectric project was being carried out.

In this context, it was revealed that when studying a river basin, it is very important to consider all the physical and natural aspects of its space, but not just the course of a river, because its area is made up of all its tributaries and the main river. Above all, the spatial organisation carried out by man as he relates to the social environment and nature must be taken into account throughout the river's domain, from the point of interference to its mouth, because the way in which events have taken place in the study area and throughout the Lower São Francisco ratifies this assertion.

It is well known that those affected by dams are claiming their rights in relation to their losses in the area that has been occupied by the water mirror of the hydroelectric dams. In the case of the São Francisco River, this issue has been much discussed since the Três Marias Dam was built and, above all, with the advent and realisation of the Sobradinho Dam, as well as all the others along the riverbed, which is synonymous with many vital meanings, as well as being a miracle-worker, among others, for its riverside inhabitants. As you can see, the work went beyond this space.

With all this said, as the various shortcomings regarding the Xingó EIA/Rima have already been highlighted, it is now time to draw attention to the core of this study, the third chapter, because it focuses on a very relevant result achieved by its main objective. This result shows the inefficiency of the EIA/Rima in general when it comes to studying the impacts of hydroelectric projects, especially those with a power generation capacity of more than 10 MW. The study took this direction because it was possible to observe that the impacts were triggered from the Xingó Dam wall to the mouth of the São Francisco River because its hydrographic basin, to this extent, is not considered to be of direct influence.

Now, if Enge-RIO took the view that from the Xingó wall to the mouth of the river, its hydrographic basin should not be considered as an area of direct influence, this company goes against this study, since Conama Resolution 01/86 states that the hydrographic basin in which the hydroelectric project will take place must be taken into account in any case in the EIA/Rima.

As a result, this result also shows that the EIA/Rima is an inadequate instrument for the study of impact forecasting, especially in this case, if what is stated in article 5, item III, of Conama Resolution 01/86 is not in line with what this study has revealed. On the other hand, if the inadequacy of the EIA/Rima is not confirmed in this case, at least believes that Enge-RIO failed to comply with the Conama Resolutions consulted.

This study draws attention not only to the case discussed here, but also to the environmental issue as a whole, the environment, the space where the relationship between man and nature can exist. What has been discussed here does not mean that man cannot generate energy from hydropower, but that energy generation from hydropower takes place responsibly. In other words, that at least the impacts caused by this action are mitigated through the application of effective policies. In addition, the managers of the units that generate this type of energy, together with the responsible authorities, should take into account what science reveals about this energy model and its consequences when poorly managed.

Compared to this model in terms of environmental viability, wind power has its advantages, since a wind power plant can be installed and generating energy in just six months. A hydroelectric plant, on the other hand, can spend an average of five years under construction before the turbines come on stream.

The cost of generating energy from wind power is high, but the cost of hydroelectric power was not low when this source began to be exploited. Nowadays, it can be said that wind power in Brazil does not have as many impacts as hydroelectric power. Thus, between the two sources, it is more viable to generate energy from the former than the latter.

This emphasises the importance of the Brazilian state

has to accelerate investments in the wind energy sector, both in science and technology, especially by advancing relevant research. In this way, it is hoped that Brazil will not be left behind, since it has great potential in this energy matrix, and that it will not have to import patented technology from abroad, since there are well advanced European countries aiming to market this energy source.

The state, the nation's supreme political-federative body, has superior power both in terms of the applicability and direction of policies for all the state sectors of the federation.

Thus, although the various hydroelectric dams installed on the São Francisco riverbed, from Sobradinho to Xingó, are in line with the energy sector's policy, it is important for the state to urgently review the operating policies of these projects with a view to correcting the impacts. Few policies to mitigate the impacts caused by the realisation and operation of these hydroelectric plants have been implemented in relation to their wide area of actual coverage. Not to mention that many places, such as Porto da Folha/SE, located in the region and also suffering from socio-environmental impacts, have not even had policies implemented in this regard.

The set of government actions was not applied to the entire area subject to the impacts of the Xingó HPP, but only to the area of influence considered by the plant's EIA/Rima. From this direction, it is believed that there is certainly a range of mistakes, as in the case of the installation and operation process of this hydroelectric power station, because the EIA should have been submitted to IBAMA for approval in the first place, but it was not.

As for the policies directed by certain state bodies, despite the failures of many, some deserve to be highlighted for their importance.

It was noted that impact mitigation policies have been applied in certain areas far from Porto da Folha/SE, but they do not refer to the impacts caused by Xingó, but by the Paulo Afonso/BA hydroelectric complex.

On the other hand, the commitment to environmental compensation with the creation of a Conservation Unit by a certain body, which according to the first reliable source exists, has not been observed, so much so that the region is facing a process of environmental degradation.

The creation of the Xingó Canal is an important water resource policy, provided that its EIA/Rima is duly approved by the competent state bodies and discussed by society in general and scientists in the field. However, until the conclusion of the study, this action had not actually been carried out. What's more, all the settlement projects in the area are not fully effective in terms of the policies they implement, and several are literally without assistance in this regard.

One fact that caught my attention was the policy adopted by one of the certain public bodies when it expropriated the fertile plain of the former Ilha do Ouro farm and left it as it is, without political-agrarian support. It is therefore difficult to believe in the process of democratisation due to the lack of effective policies on this site and on the entire area of the high sequeiros that need water to carry out agricultural activities.

It emerged that although the sharecroppers were subject to the wealthy landowners of the area during the rice cycle, they usually had this cereal to eat all year round. Nowadays, on the other hand, they don't get it in the same way, nor do they get the other sources of food that many of these farmers in the same conditions had in abundance. What's even worse is that they haven't reached the promised land, as this requires an abundance of water. In this context, the application of policies at that time would have been efficient and effective if there had been supervisory measures to prevent inhumane, illegal and unjust activities such as those that took place.

The way in which the state has behaved since the Xingó hydroelectric project to this day, in terms of directing and applying policies to mitigate impacts, highlights what is believed to be a historical debt to the place, in the sense of the absence of these policies.

This is part of a prospective vision of the regenerative capacity of the local ecosystem. Thus, if the state revises operating policies such as controlling the flow of hydroelectric dams along the course of the São Francisco, from Xingó to Três Marias, it is possible to regenerate the entire ecosystem that existed during the rice cycle. To achieve this, the policies needed to realise this process of environmental recovery don't stop there; the state needs to act by implementing many policies in addition to those mentioned above. Among the many other important policies to be implemented effectively, the educational, environmental and agricultural policies that are viable for the environment deserve to be emphasised. With regard to the former, it is important to implement a programme aimed at raising knowledge and awareness about the environment. It is also important to train students so that, together with environmental technicians, they can take the same educational approach to the subject to both urban and rural communities. It is important that this policy reaches, especially, the peasant farmers who live and depend on agriculture.

Environmental policy in the area has an important role to play in restoring the area where the tropical hydrophilous floodplain forest was extinguished. This role, among other things, can be confirmed by planting native vegetation seedlings in the appropriate areas, as directed by competent technicians. It is essential to create favourable laws to protect this environment once it has been recovered, so that it is not devastated again.

Finally, the third policy focuses on the agricultural sector, but this government action should promote the intertwining of agricultural development and environmental preservation, so that this development does not lead to the degradation of nature. It is hoped that the state will subsidise small farmers in the region, both financially and with technical guidance through its sectoral bodies. In extreme cases and according to the need for technical guidance, these policies should also be extended to medium and large landowners. In addition, the environmental laws created should address this context. Technicians in this sector need to be able to guide farmers so that they can promote the agricultural development they practise while preserving the environment.

However, if the state doesn't act in this direction, the ecosystem, which during the rice cycle was an enclave in the sertão of Sergipe, tends to become increasingly vulnerable to the desertification process that is spreading in the northeast. In this sense, it is now possible to see small concentrations of salt in the soil of the extensive river plain, which used to be unusual. This is therefore a strong indication of this process of environmental degradation.

Based on all the information obtained, together with all the data collected and contextualised, the hypothesis initially raised was confirmed. The socio-environmental impacts on the banks of the São Francisco River, in Porto da Folha/SE, were caused mainly by the construction and operation of the Xingó Hydroelectric Power Plant. Even though other dams on the riverbed indirectly contributed to the episode, their impacts were not significant compared to those of the Xingó Hydroelectric Power Plant. The socio-environmental impacts of the Xingó hydroelectric dam only became apparent in the area after it came into existence.

It is essential to state that if all the impacts identified here had been avoided, the physical and natural factors of the place would certainly not have been added to the effects of climate change caused by global warming. All because, among other factors, plant biomass has the natural function of balancing the temperature of the environment in which it is inserted. Thus, due to the de-characterisation of the enclave that existed, the ambient temperature is no longer the same as it was at that time, given the scenario found today. This scenario has already been characterised, so it is part of the factors that cause global warming, for example: the presence of bare land and the reduction of the water mirror also contribute to the increase in temperature. In this context, among so many other circumstances facing the complex issue of global warming, the subject was tackled with the aim of highlighting ways to achieve the long-awaited sustainable development.

On top of everything else, the worst is yet to come: the negative impacts that will henceforth arise after the transposition of the river is carried out, which will be added to those that already exist, as already stated. The following should be emphasised in this regard: sooner or later, serious studies aimed at understanding the causes and effects of this work (transposition) will surely reveal immeasurable and perhaps irreversible negative impacts. In saying that the transposition will make the impacts even worse, it is enough to measure them against the sum of the hydroelectric projects that have resulted in a drastic reduction in the river's flow. It is well known that reservoirs tend to reserve as much water as possible at certain times to prevent them from reaching minimum levels or for other political reasons. Certainly, the reservoirs immediately adjacent to the points where the canals come out (the North (Cabrobó/PE) and East (Petrolândia/PE) axes) will have even greater control over this situation, which will harm the lower course of the river. The latter is already suffering from the influence of the sea due to significant changes in its flow.

If, on the one hand, there are those who say that the transposition will serve life, bringing water to the needy people of isolated and poor areas of this natural resource, on the other hand, there are those who say that you have to wait for everything to happen to believe the doubt that exists in this regard, because the written and spoken press shows that the canals are for irrigation. Finally, the reason for this statement is that this study has shown that many municipalities on the banks of the flowing São Francisco River have several times been declared in a state of public calamity due to a lack of water for their people.

This study revealed the following:

The search for solutions to environmental issues is fundamental. It is therefore essential that man adapts to the environment and not the other way round. It's like the law of gravity, which man does not fully and naturally defy. And so it is with the environment: man cannot make it adapt to being, but living beings can adapt to nature.

BIBLIOGRAPHICAL REFERENCES[22]

AB'SABER, Aziz Nacib. **The domains of nature in Brazil**: landscape potential.

[22] According to:

São Paulo: Ateliê Editorial, 2003. 160 p.

_____. Conceptual bases and the role of knowledge in impact forecasting. In: AB'SABER, Aziz Nacib; PLANTENBERG, Clarita Muller (eds.). **Impact forecasting**: environmental impact assessment in the East, West and South: experiences in Brazil, Russia and Germany. 2ª edition. São Paulo: Edusp, 2006. p. 27-49.

AB'SABER, Aziz Nacib & PLANTENBERG, Clarita Muller (org). **Impact forecasting**. 2nd edition. São Paulo: Edusp, 2006. 573 p.

ALMEIDA, Lúcia M. Alves de; RIGOLIN, Tércio B. The environmental question: nature, society and technology. In: _______________ . **Geography**: series - new high school education. São Paulo: ática publishing house, 2002. p. 07-186.

ALVES, Aloar Café. Environmental analysis from a legal point of view. In: TANK, Sâmia Maria. (org.). **Análise ambiental**: uma visão multidisciplinar. 2a edição. São Paulo: Unesp, 1995. p. 65-71.

ANDRADE, Manuel Correia de. **A terra e o homem do Nordeste**: contribuição ao estudo da questão agrária no Nordeste. 7th edition. São Paulo: Cortez publishing house, 2004. 318 p.

BASTOS, Eduardo Alves; FONSECA, Vânia. The economic dimension. In: FONSECA, Vânia; BASTOS, Eduardo Alves (coords.). **Sertão do baixo São Francisco sergipano**: bacia hidrográfica como unidade de estudo. Aracaju: Codevasf/UFS/Cnpq, 1997. p. 10-16.

BEZERRA, Felte. **From the land**. 1938. 127 f. Thesis presented to the Congregation of the Sergipe Athenaeum, to compete for the post of Professor of Geography. Aracaju, 1938.

BRAZIL. National Environmental Council - Brazilian Institute of the Environment and Renewable Natural Resources (Conama/Ibama). For the purposes of this Resolution, environmental impact is considered... Resolution

Conama, No. 01, 23 January 1986. **Resolutions,** Brasilia, February 1986.

_____. Provides for the environmental licensing of works in the electricity generation sector. Conama Resolution No. 6, 16 September 1987. **Resolutions**, Brasilia, October 1987.

BRAZIL. Constitution (1988). **The Constitution of Brazil 1988**: compared and commented. São Paulo: Price Waterhouse, 1989, 1028 p.

CARVALHO, Marcos de. **What is nature**. 2nd edition, 2nd reprint· São Paulo:

BRAZILIAN ASSOCIATION OF TECHNICAL STANDARDS. NBR 6023: information and documentation: references: elaboration. Rio de Janeiro, 2002.

brasiliense, 2003. 85 p.

CÉPÈDE, M.; HOUTART, F.; GOND, L. Las subsistencias. In: ________. **The world population and the means of subsistence**. Coleccion "El Mundo y Los Hombres" n° 14. Barcelona - 11: Nova Terra publishing house, 1967. p. 235-405.

COMPANHIA HIDRO ELÉTRICA DO SÃO FRANCISCO - Chesf. **Inflow and outflow data for the Itaparica, Xingó and Sobradinho reservoirs** [personal message]. Message received by <regnaldogouveia@bol.com.br > on 27 March 2007.

CONTI, José Bueno. **Climate and the environment**. 4th edition. São Paulo: Atual, 1998. 88 p.

COSTA, Reginaldo Brito da. Agroforestry systems (SAFs): sustainable production alternatives for family farming. In: SILVA, Medson Janer da (org.). **Métodos e culturas alternativas na agricultura familiar**. Campo Grande: Editora Ucdb, 2003. p. 113-124.

CUNHA, José Carlos Santos. **Traditional farming techniques**: - Social/environmental efficiency in the semi-arid region of Sergipe. 1997. 154 f. Master's dissertation - Npgeo/UFS, Aracaju, 1997.

SERGIPE AGRICULTURAL DEVELOPMENT COMPANY (Emdagro/SE). Porto da Folha local unit. **Rainfall data - 1979-2006**. Porto da Folha, 2007. 1 diskette, 3 ^ in. Word for Windows 7.0.

Enge-RIO - ENGENHARIA E CONSULTORIA S.A. **Xingó hydroelectric power station**: Environmental Impact Study - EIA. Presentation. Rio de Janeiro: [s.n.]. 1993. 34 p.

. **Xingó Hydroelectric Power Plant**: Environmental Impact Study - EIA. Volume I - descriptive memorial of the project. Rio de Janeiro: [s.n.]. 1993. 113 p.

_____. **Xingó Hydroelectric Power Plant**: Environmental Impact Study - EIA. Volume II - environmental diagnosis, volume 01 - physical environment. Rio de Janeiro: [s.n.]. 1993. 282 p.

_____. **Xingó Hydroelectric Power Plant**: Environmental Impact Study - EIA. Volume II - environmental diagnosis, volume 02 - biotic environment. Rio de Janeiro: [s.n.]. 1993. 201 p.

_____. **Xingó Hydroelectric Power Plant**: Environmental Impact Study - EIA. Volume II - environmental diagnosis, volume 03 - anthropic environment. Rio de Janeiro: [s.n.]. 1993. 208 p.

_____. **Xingó Hydroelectric Power Plant**: Environmental Impact Study - EIA. Volume III - environmental prognosis. Rio de Janeiro: [s.n.]. 1993. 86 p.

_____. **Xingó Hydroelectric Power Plant**: Environmental Impact Study - EIA. Volume IV - environmental programmes. Rio de Janeiro: [s.n.]. 1993. 64 p.

_____. **Environmental Impact Report - RIMA**: Xingó Hydroelectric Power Plant - UHE XINGÓ. Rio de Janeiro, [s.n.], [1993 or 1994]. 108 p.

_____. General aspects of the project. In: _________________ . **Xingó Hydroelectric Power Plant. Basic Environmental Project - PBA**: aquatic fauna management and conservation programme. Rio de Janeiro: [s.n.]. 1994. 1-10 p.

_____. Aquatic fauna management and conservation project. In: _______. **Xingó Hydroelectric** Power Station. **Basic Environmental Project - PBA**: aquatic fauna management and conservation project. Rio de Janeiro: [s.n.]. 1994. 11-62 p.

FERREIRA, J. P. **Enciclopédia dos municípios brasileiros**. XIX vol., Rio de Janeiro: Ibge, 1959.

BRAZILIAN INSTITUTE OF GEOGRAPHY AND STATISTICS - IBGE. **Agricultural** census: IX general census of Brazil - 1980 - Sergipe. Volume 2 - volume 3 - number 14. Rio de Janeiro: Ibge, 1983.

GONÇALVES, Neyde Maria Santos. **Rainfall impacts and disorganisation of urban space in Salvador/BA**. 1992. 268 f. Thesis (Doctorate in physical geography) - Dpgeo - Faculty of Philosophy, Letters and Human Sciences of the University of São Paulo (Fflch/USP.), São Paulo, 1992.

BRAZILIAN INSTITUTE OF GEOGRAPHY AND STATISTICS. **Economic censuses**: agricultural, industrial, commercial and services - general census of Brazil (1st September 1940), regional series - part XI - Sergipe. Rio de Janeiro: Ibge, 1952.

_____. **Agricultural production**: broken down by municipality - 1953 - Sergipe. Rio de January: Ibge, 1954.

_____. **1960 agricultural** census: 7th general census of Brazil - regional series - Alagoas - Sergipe. Volume II - Volume VII - Part 1ª. Rio de Janeiro: Ibge, 1960.

_____. **Agricultural production**: broken down by municipality - 1974-1990 - Sergipe. Rio de Janeiro: Ibge, (17 vols.).

_____. **1980 Agricultural Census**: IX General Census of Brazil - Sergipe. Volume II - volume III - no. 14. Rio de Janeiro: Ibge, 1983.

_____. **Diagnosis of the environmental quality of the São Francisco River basin, western Bahia and Sobradinho sub-basins**: Geosciences Studies and Research Series - No. 02. Rio de Janeiro: Ibge, 1994. p. 21-104.

. **Municipal Agricultural Production - PAM, Sidra**. Available at:< http://www.ibge.gov.br>. Accessed on: 04 April 2007.

LAMARCHE, Hugues (corrd.) Productive logics. In: ___________ . **A Family Farming**: International Comparison - From Myth to Reality. Translation by Frédéric Bazin. Volume II. Campinas: Unicamp publishing house, 1998. p. 61-88.

_____. Towards a theory of family farming. _________. **Family Farming**: International Comparison - From Myth to Reality. Translation by Frédéric Bazin. Volume II. Campinas: Unicamp publishing house, 1998. p. 303-336.

_____. **2010 Population Count**: Census population, by municipality - Sergipe - **2010** _2010. Available at: <http://www.censo2010.ibge.gov.br>. Accessed on: 20 November 2010.

LENCIONI, Sandra. **Region and Geography**. 1st edition. São Paulo: Edusp. 2003. 224 p.

LIMA, Dalmo M. de Albuquerque; WILKINSON, John et al. (eds.). Cnpq's initiatives in ST&I to support family farming and agrarian reform settlements. In: ______________________________ . **Innovation in the traditions of family farming**. Parallel 15. Brasília: Cnpq, 2002. 400 p.

LIMA, J. Hunaldo. **Várzea da Ilha do Ouro**: hypsometry and Inundable Areas (A. I.) - 1987. Aracaju: [s.n.], 2007. 1 map, colour, 9.18 cm x 10.37 cm. Scale 1:30,000.

_____. Primitive vegetation and flood areas: Municipality of Porto da Folha/SE - 1979. Aracaju: [s.n.], 2007. 1 map, colour, 10.37 cm x 10.48 cm. Scale 1:300.000.

LOCAL. In: OLIVEIRA dicionário cartográfico. 2a ed. Rio de Janeiro: Ibge, 1983. p. 377.

PLACE. In: OLIVEIRA dicionário cartográfico. 2nd ed. Rio de Janeiro: Ibge, 1983. p. 380.

MESQUITA, Elpídio de. **History of the São Francisco River**. 17. Rio de Janeiro: [s.n.], [192?]. 327 p.

MILARÉ, Édis. Procedure and content. In: MILARÉ, Édis & BENJAMIN, Antonio Herman V. **Estudo prévio de impacto ambiental - teoria, prática e legislação**. São Paulo: Revista dos Tribunais, 1993. p. 166.

. Prior environmental impact assessment in Brazil. In: AB'SABER, Aziz Nacib; PLANTENBERG, Clarita Muller (eds.). **Forecasting impacts**: environmental impact studies in the East, West and South: experiences in Brazil, Russia and Germany. 2nd edition. São Paulo: Edusp, 2006. p. 51-83.

MINISTRY OF AGRICULTURE, INDUSTRY AND COMMERCE - GENERAL DIRECTORATE OF STATISTICS. **Census of Brazil - 1920**: Sergipe. Volume III. Rio de Janeiro: 1924.

MINISTRY OF NATIONAL INTEGRATION - COMPANHIA DE DESENVOLVIMENTO DOS VALES DO SÃO FRANCISCO E DO PARNAÍBA (Codevasf) - 4th REGIONAL SUPERINTENDENCY. **Map of the sub-basins of the Lower São Francisco Sergipano - Pedology**. Aracaju: Codevasf, 2000. 1 colour map. Scale 1: 100,000.

MINISTRY OF THE INTERIOR - NORTHEAST DEVELOPMENT SUPERINTENDENCE. **Topographical map of Pão de Açúcar/AL**. Recife: 1973, 1 sheet (SC.24-X-D-IV). Scale 1:100,000.

EXTRAORDINARY MINISTRY OF LAND POLICY - NATIONAL INSTITUTE FOR COLONISATION AND AGRARIAN REFORM - RURAL REGISTRATION DIVISION - SR-23/C. **Ilha do Ouro settlement**: general plan. Aracaju, 1998, 1 plan. Scale: 1:10,000.

MONOSOWSKI, Elizabeth. The Sertão will turn into the sea... Environmental assessment and management at the Tucuruí dam, Amazonia. In: AB'SABER, Aziz Nacib; PLANTENBERG, Clarita Muller (eds.). **Impact forecasting**: environmental impact assessment in the East, West and South: experiences in Brazil, Russia and Germany. 2ª edition. São Paulo: Edusp, 2006. p. 123-141.

MONTEIRO, Carlos A. de Figueiredo. **Aspectos geográficos do Baixo São Francisco**. São Paulo, Association of Brazilian Geographers (Avulso n° 5), 1962. 95 p.

MOTT, Luiz R. B. **Sergipe Del Rey**: population, economy and society. Aracaju: Fundesc, 1986. 204 p.

NASCIMENTO, Miguel do; FÁBIO; LÁZARO. **Monitoring Report - Department of Environmental Assessment and Monitoring - DEAMOAM**. Aracaju: ADEMA/SE, 2007. 07 p. (N° 09/2007).

OLIVEIRA, Ariovaldo de. **Peasant agriculture in Brazil**. 4th edition. São Paulo: Contexto, 2001. 164 p.

GRANT OF THE TITLE OF CITIZEN OF RIO-CLARENSE TO TROPPMAIR, H., 1985, Rio Claro. **Text for critical reading - Man in the face of the environment**: "Why bequeath to those to whom we have given our lives a sombre future, a world of hatred, hunger, pollution and hopelessness?" Rio Claro: [s.n.], May 1985.

PRADO, Ivo do. **A Capitania de Sergipe e suas ouvidorias**: memórias sobre questões **de** limites. Rio de Janeiro: Papelaria Brasil, 1919. 411 p.

REBOUÇAS, Aldo da Cunha. Fresh water in the world and in Brazil. In:

REBOUÇAS, Aldo da Cunha; BRAGA, Benedito; TUNDISI, José Galízia (eds.). **Fresh waters in Brazil**: ecological capital, use and conservation. São Paulo: Escrituras, 1999. p. 1-38.

ROLSTON III, Holmes. Philosophy of nature and ecology. In: CHEDIAK, Karla; VIEIRA, Antônio Augusto P. (eds.). **Themes in the philosophy of nature**. Rio de Janeiro: Uerj, 2004. p. 299-341.

RODRIGUES, Cleide; ADAMI, Samuel. Fundamental techniques for the study of river basins. In: VENTURI, Luiz A. Bittar. **Practising geography**: field and laboratory techniques. São Paulo: Oficina de Textos, 2005. p. 147-166.

ROSS, Jurandir L. Sanches (org.) **Geography of Brazil**. 5ª edition rev. e ampl. São Paulo: Edusp, 2005. 549 p.

SACHS, Ignacy. From Economic Growth to Ecodevelopment. In: VIEIRA, Paulo Freire, et al. (org.). **Desenvolvimento e Meio Ambiente no Brasil**: a Contribuição de Ignacy Sachs. Porto Alegre: Pallotti; Florianópolis: Aped, 1998, p. 161-163.

SÁNCHEZ, Luis E. Concepts and Definitions. In: ______ . **Evaluation of Environmental Impact**: concepts and methods. São Paulo: Oficina de Textos, 2006. Chapter 01. p. 17-43.

SANTOS, Aldeci Figueiredo; ANDRADE, José Augusto. **Delimitation and regionalisation of semi-arid Brazil**. Aracaju: UFS, 1992. 232 p.

SANTOS, Maria Genisete G. dos. Interview given to Regnaldo Gouveia dos Santos. Porto da Folha/SE, 18th January 2007.

SANTOS, Nelson Martins dos. Interview with Regnaldo Gouveia dos Santos. Porto da Folha/SE, 18th January 2007.

SANTOS, Regnaldo Gouveia dos. Socio-environmental impacts on the banks of the São Francisco River: a result of the lack of consideration of the Area of Real Influence. **Revista Geousp: Espaço e Tempo**, São Paulo, Special Edition, p. 81-91, December 2009.

SANTOS, Thereza Christina Carvalho; CÂMARA, João Batista Drumond (eds.). The study of water resources. In: ________________ . **Geo Brasil 2002**: perspectivas do meio ambiente no Brasil, Brasília: Ibama, MMA. 2002. p. 78- 156.

SARTORI, Giovani. What is Politics? In: _________. **Politics**: Logic and method in the social sciences. Translation Sérgio Bath. 2ª edition. Brasília: Editora Universidade de Brasília, 1997. p. 157-174.

SENE, Eustáquio de; MOREIRA, João Carlos. **General and Brazilian geography**: geographical space and globalisation. São Paulo: Scipione, 2000. 503 p.

SERGIPE. SUVALE/ANCARSE. **Action plan for the humid valleys of the Lower São Francisco in Sergipe**. Aracaju: Ancarse, 1972.

SILVA, J. F. Graziano da (coord.). **Estrutura agrária e produção de subsistência na agricultura brasileira**. São Paulo: Editora Hucitec, 1978. 268 p.

SILVA, Luciano Meneses Cardoso da. Sustainable Reservoir Management. In: THEODORO, Suzi Huff (org). **Conflicts and sustainable use of natural resources**. Rio de Janeiro: Garamond Ltda, 2002. p. 29-49.

SOUZA, Marcos Antonio de. **Memória sobre a Capitania** de **Sergipe**. 2nd edition. Aracaju: IBGE, 1944. 44 p.

SUPERINTENDENCE OF NORTHEAST DEVELOPMENT / NORTHEAST DEVELOPMENT AGENCY - SUDENE/ADENE. **Data on droughts and floods in the Northeast** [personal message]. Message received by <regnaldogouveia@bol.com.br > on 23 November 2005.

TEIXEIRA, Maria Gracinda et al. Analysis of environmental impact reports for large hydroelectric dams in Brazil. In: AB'SABER, Aziz Nacib; PLANTENBERG, Clarita Muller (eds.). **Impact forecasting**: environmental impact assessment in the East, West and South: experiences in Brazil, Russia and Germany. 2nd edition. São Paulo: Edusp, 2006. p. 163-186.

THÉRY, Hervé; MELLO, Neli Aparecida de. The environment and its management. In: . **Atlas of Brazil**: disparities and dynamics of the territory. São Paulo: Edusp, 2005. Chapter 03. p. 60-88.

TREPL, Ludwig. What can "environmental impact" mean? In: AB'SABER, Aziz Nacib; PLANTENBERG, Clarita Muller (org). **Impact forecasting**. 2nd edition. São Paulo: Edusp, 2006. p. 329-350.

VALVERDE, Orlando. Regional division of the São Francisco valley. **Revista brasileira de geografia**, Brazil, year VI, 2nd issue, p. 179-218, April-June, 1944.

VARGAS, Maria Augusta Mundim. **Regional Development in Question**: The Baixo São Francisco Revisited. São Cristóvão: Npgeo/UFS, 1999.

WEBER, Max. Science and Politics. In: _______. **Essays on sociology**. Translation by Waltensir Dutra. 5ª edition. Rio de Janeiro: Zahar editores, 1982. p. 97-153.

WOORTMANN, Ellen F. The sitiantes of the Northeast. In: __. **Heirs, relatives and compadres**. São Paulo: Hucitec, 1995. p. 217-239.

ANNEXES

Annex A - Old Charter of Sergipe Del Rei

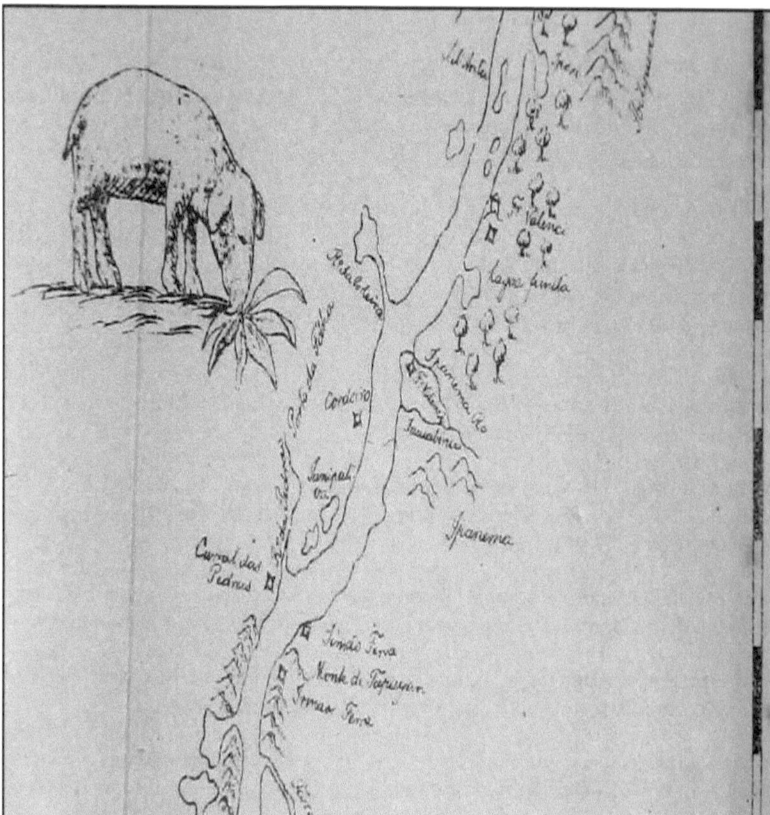

Source: A Capitania de Sergipe e suas ouvidorias: memórias sobre questões de limites - 1919.

Annex B - Map of the State of Sergipe and part of the State of Bahia

Source: A Capitania de Sergipe e suas ouvidorias: memórias sobre questões de limites - 1919.

Annex C - Letter from the District of Sergipe Del Rei

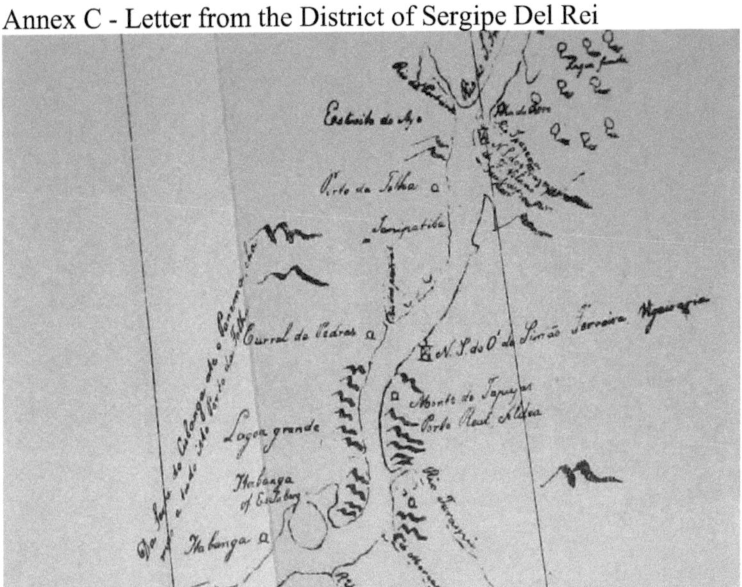

Source: A Capitania de Sergipe e suas ouvidorias: memórias sobre questões de limites - 1919.

Annex D - Part of the topographical map of Pão de Açúcar/AL

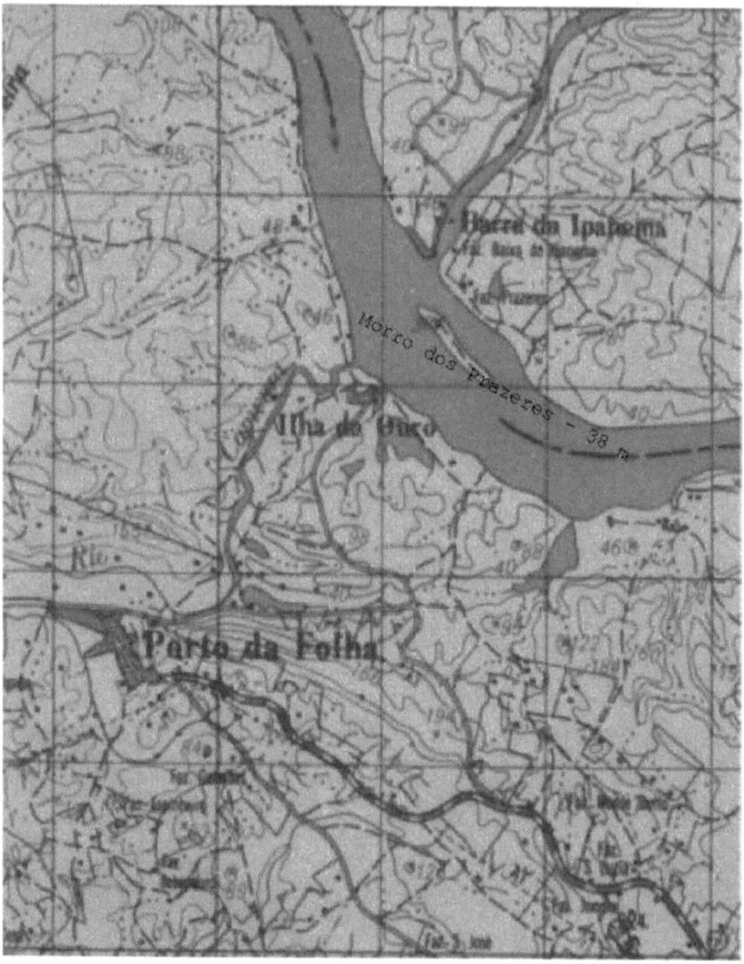

Source: MINISTÉRIO DO INTERIOR - SUPERINTENDÊNCIA DO DESENVOLVIMENTO DO NORDESTE. Topographical map of Pão de Açúcar/AL - 1973.

Annex E - General rules for Environmental Licensing of hydroelectric plants according to Conama Resolutions 01/86 and 06/87.

ETA PAS	RESOLUTION Conama N° 01/86	RESOLUTION Conama N° 06/87	HYDROELECTRIC
Iª	It will depend on the EIA/Rima and its approval by the competent body and Ibama, after being submitted to them.		With a potential of more than 10MW
2ª	0 The EIA must define the geographical area directly or indirectly affected by the impacts of the project.		In this case, the hydrographic basin in which the HPP is located is considered
3ª		There needs to be an understanding between state agencies	Projects involving more than one state
4ª		The LP must be applied for	At the start of the feasibility study
5ª		The LI must be requested	Before the construction tender
6ª		The LO must be obtained	Before the dam was closed

Annex F - Conama Resolution 001 of 23 January 1986

RESOLUÇÃO CONAMA nº 1, de 23 de janeiro de 1986
Publicada no DOU, de 17 de fevereiro de 1986, Seção 1, páginas 2548-2549

Correlações:
- Alterada pela Resolução nº 11/86 (alterado o art. 2º)
- Alterada pela Resolução nº 5/87 (acrescentado o inciso XVIII)
- Alterada pela Resolução nº 237/97 (revogados os art. 3º e 7º)

Dispõe sobre critérios básicos e diretrizes gerais para a avaliação de impacto ambiental

O CONSELHO NACIONAL DO MEIO AMBIENTE - CONAMA, no uso das atribuições que lhe confere o artigo 48 do Decreto nº 88.351, de 1º de junho de 1983, [156]para efetivo exercício das responsabilidades que lhe são atribuídas pelo artigo 18 do mesmo decreto, e

Considerando a necessidade de se estabelecerem as definições, as responsabilidades, os critérios básicos e as diretrizes gerais para uso e implementação da Avaliação de Impacto Ambiental como um dos instrumentos da Política Nacional do Meio Ambiente, resolve:

Art. 1º Para efeito desta Resolução, considera-se impacto ambiental qualquer alteração das propriedades físicas, químicas e biológicas do meio ambiente, causada por qualquer forma de matéria ou energia resultante das atividades humanas que, direta ou indiretamente, afetam:

I - a saúde, a segurança e o bem-estar da população;
II - as atividades sociais e econômicas;
III - a biota;
IV - as condições estéticas e sanitárias do meio ambiente;
V - a qualidade dos recursos ambientais.

Art. 2º Dependerá de elaboração de estudo de impacto ambiental e respectivo relatório de impacto ambiental - RIMA, a serem submetidos à aprovação do órgão estadual competente, e da Secretaria Especial do Meio Ambiente - SEMA[157] em caráter supletivo, o licenciamento de atividades modificadoras do meio ambiente, tais como:

I - Estradas de rodagem com duas ou mais faixas de rolamento;
II - Ferrovias;
III - Portos e terminais de minério, petróleo e produtos químicos;
IV - Aeroportos, conforme definidos pelo inciso 1, artigo 48, do Decreto-Lei nº 32, de 18 de setembro de 1966[158];
V - Oleodutos, gasodutos, minerodutos, troncos coletores e emissários de esgotos sanitários;
VI - Linhas de transmissão de energia elétrica, acima de 230KV;
VII - Obras hidráulicas para exploração de recursos hídricos, tais como: barragem[159] para fins hidrelétricos, acima de 10MW, de saneamento ou de irrigação, abertura de canais para navegação, drenagem e irrigação, retificação de cursos d'água, abertura de barras e embocaduras, transposição de bacias, diques;
VIII - Extração de combustível fóssil (petróleo, xisto, carvão);
IX - Extração de minério, inclusive os da classe II, definidas no Código de Mineração;
X - Aterros sanitários, processamento e destino final de resíduos tóxicos ou perigosos;

---

156 Decreto revogado pelo Decreto nº 99.274, de 6 de junho de 1990.
157 A Secretaria Especial do Meio Ambiente – SEMA, vinculada ao Ministério do Interior, foi extinta pela Lei nº 7.735, de 22 de fevereiro de 1989, que criou o Instituto Brasileiro do Meio Ambiente e dos Recursos Naturais Renováveis – IBAMA. As atribuições em matéria ambiental são atualmente do Ministério do Meio Ambiente.
158 Decreto-Lei revogado pela Lei nº 7.565, de 19 de dezembro de 1986.
159 Retificado no Boletim de Serviço do MIN, de 7 de março de 1986

XI - Usinas de geração de eletricidade, qualquer que seja a fonte de energia primária, acima de 10MW;

XII - Complexo e unidades industriais e agro-industriais (petroquímicos, siderúrgicos, cloroquímicos, destilarias de álcool, hulha, extração e cultivo de recursos ~~hídricos~~ hidróbios?)[160];

XIII - Distritos industriais e zonas estritamente industriais - ZEI;

XIV - Exploração econômica de madeira ou de lenha, em áreas acima de 100 hectares ou menores, quando atingir áreas significativas em termos percentuais ou de importância do ponto de vista ambiental;

XV - Projetos urbanísticos, acima de 100 ha ou em áreas consideradas de relevante interesse ambiental a critério da SEMA e dos órgãos ~~municipais e estaduais competentes~~ estaduais ou municipais[1];

~~XVI - Qualquer atividade que utilizar carvão vegetal, em quantidade superior a dez toneladas por dia.~~

XVI - Qualquer atividade que utilizar carvão vegetal, derivados ou produtos similares, em quantidade superior a dez toneladas por dia. (*nova redação dada pela Resolução nº 11/86*)

XVII - Projetos Agropecuários que contemplem áreas acima de 1.000 ha. ou menores, neste caso, quando se tratar de áreas significativas em termos percentuais ou de importância do ponto de vista ambiental, inclusive nas áreas de proteção ambiental. (*inciso acrescentado pela Resolução nº 11/86*)

XVIII - Empreendimentso potencialmente lesivos ao patrimônio espeleológico nacional. (*inciso acrescentado pela Resolução nº 5/87*)

~~Art. 3º Dependerá de elaboração de estudo de impacto ambiental e respectivo RIMA, a serem submetidos à aprovação da SEMA, o licenciamento de atividades que, por lei, seja de competência federal.~~ (*Revogado pela Resolução nº 237/97*)

Art. 4º Os órgãos ambientais competentes e os órgãos setoriais do SISNAMA deverão compatibilizar os processos de licenciamento com as etapas de planejamento e implantação das atividades modificadoras do meio ambiente, respeitados os critérios e diretrizes estabelecidos por esta Resolução e tendo por base a natureza o porte e as peculiaridades de cada atividade.

Art. 5º O estudo de impacto ambiental, além de atender à legislação, em especial os princípios e objetivos expressos na Lei de Política Nacional do Meio Ambiente, obedecerá às seguintes diretrizes gerais:

I - Contemplar todas as alternativas tecnológicas e de localização do projeto, confrontando-as com a hipótese de não execução do projeto;

II - Identificar e avaliar sistematicamente os impactos ambientais gerados nas fases de implantação e operação da atividade;

III - Definir os limites da área geográfica a ser direta ou indiretamente afetada pelos impactos, denominada área de influência do projeto, considerando, em todos os casos, a bacia hidrográfica na qual se localiza;

IV - Considerar os planos e programas governamentais, propostos e em implantação na área de influência do projeto, e sua compatibilidade.

Parágrafo único. Ao determinar a execução do estudo de impacto ambiental o órgão estadual competente, ou a SEMA ou, no que couber ao Município [161], fixará as diretrizes adicionais que, pelas peculiaridades do projeto e características ambientais da área, forem julgadas necessárias, inclusive os prazos para conclusão e análise dos estudos.

Art. 6º O estudo de impacto ambiental desenvolverá, no mínimo, as seguintes atividades técnicas:

160 Retificado no Boletim de Serviço do MIN, de 7 de março de 1986

161 Retificado no Boletim de Serviço do MIN, de 7 de março de 1986

I - Diagnóstico ambiental da área de influência do projeto completa descrição e análise dos recursos ambientais e suas interações, tal como existem, de modo a caracterizar a situação ambiental da área, antes da implantação do projeto, considerando:

a) o meio físico - o subsolo, as águas, o ar e o clima, destacando os recursos minerais, a topografia, os tipos e aptidões do solo, os corpos d'água, o regime hidrológico, as correntes marinhas, as correntes atmosféricas;

b) o meio biológico e os ecossistemas naturais - a fauna e a flora, destacando as espécies indicadoras da qualidade ambiental, de valor científico e econômico, raras e ameaçadas de extinção e as áreas de preservação permanente;

c) o meio sócio-econômico - o uso e ocupação do solo, os usos da água e a sócio-economia, destacando os sítios e monumentos arqueológicos, históricos e culturais da comunidade, as relações de dependência entre a sociedade local, os recursos ambientais e a potencial utilização futura desses recursos.

II - Análise dos impactos ambientais do projeto e de suas alternativas, através de identificação, previsão da magnitude e interpretação da importância dos prováveis impactos relevantes, discriminando: os impactos positivos e negativos (benéficos e adversos), diretos e indiretos, imediatos e a médio e longo prazos, temporários e permanentes; seu grau de reversibilidade; suas propriedades cumulativas e sinérgicas; a distribuição dos ônus e benefícios sociais.

III - Definição das medidas mitigadoras dos impactos negativos, entre elas os equipamentos de controle e sistemas de tratamento de despejos, avaliando a eficiência de cada uma delas.

IV - Elaboração do programa de acompanhamento e monitoramento dos impactos positivos e negativos, indicando os fatores e parâmetros a serem considerados.

Parágrafo único. Ao determinar a execução do estudo de impacto ambiental, o órgão estadual competente; ou a SEMA ou quando couber, o Município fornecerá as instruções adicionais que se fizerem necessárias, pelas peculiaridades do projeto e características ambientais da área.

~~Art. 7º O estudo de impacto ambiental será realizado por equipe multidisciplinar habilitada, não dependente direta ou indiretamente do proponente do projeto e que será responsável tecnicamente pelo resultados apresentados.~~ (*Revogado pela Resolução nº 237/97*)

Art. 8º Correrão por conta do proponente do projeto todas as despesas e custos referentes à realização do estudo de impacto ambiental, tais como: coleta e aquisição dos dados e informações, trabalhos e inspeções de campo, análises de laboratório, estudos técnicos e científicos e acompanhamento e monitoramento dos impactos, elaboração do RIMA e fornecimento de pelo menos 5 (cinco) cópias.

Art. 9º O relatório de impacto ambiental - RIMA refletirá as conclusões do estudo de impacto ambiental e conterá, no mínimo:

I - Os objetivos e justificativas do projeto, sua relação e compatibilidade com as políticas setoriais, planos e programas governamentais;

II - A descrição do projeto e suas alternativas tecnológicas e locacionais, especificando para cada um deles, nas fases de construção e operação a área de influência, as matérias primas, e mão-de-obra, as fontes de energia, os processos e técnicas operacionais, os prováveis efluentes, emissões, resíduos e perdas de energia, os empregos diretos e indiretos a serem gerados;

III - A síntese dos resultados dos estudos de diagnósticos ambiental da área de influência do projeto;

IV - A descrição dos prováveis impactos ambientais da implantação e operação da atividade, considerando o projeto, suas alternativas, os horizontes de tempo de incidência dos impactos e indicando os métodos, técnicas e critérios adotados para sua identificação, quantificação e interpretação;

V - A caracterização da qualidade ambiental futura da área de influência, comparando as diferentes situações da adoção do projeto e suas alternativas, bem como com a hipótese de sua não realização;

VI - A descrição do efeito esperado das medidas mitigadoras previstas em relação aos impactos negativos, mencionando aqueles que não puderem ser evitados, e o grau de alteração esperado;

VII - O programa de acompanhamento e monitoramento dos impactos;

VIII - Recomendação quanto à alternativa mais favorável (conclusões e comentários de ordem geral).

Parágrafo único. O RIMA deve ser apresentado de forma objetiva e adequada a sua compreensão. As informações devem ser traduzidas em linguagem acessível, ilustradas por mapas, cartas, quadros, gráficos e demais técnicas de comunicação visual, de modo que se possam entender as vantagens e desvantagens do projeto, bem como todas as conseqüências ambientais de sua implementação.

Art. 10. O órgão estadual competente, ou a SEMA ou, quando couber, o Município terá um prazo para se manifestar de forma conclusiva sobre o RIMA apresentado.

Parágrafo único. O prazo a que se refere o *caput* deste artigo terá o seu termo inicial na data do recebimento pelo órgão estadual competente ou pela SEMA do estudo do impacto ambiental e seu respectivo RIMA.

Art. 11. Respeitado o sigilo industrial, assim solicitando e demonstrando pelo interessado o RIMA será acessível ao público. Suas cópias permanecerão à disposição dos interessados, nos centros de documentação ou bibliotecas da SEMA e do órgão estadual de controle ambiental correspondente, inclusive durante o período de análise técnica.

§ 1º Os órgãos públicos que manifestarem interesse, ou tiverem relação direta com o projeto, receberão cópia do RIMA, para conhecimento e manifestação.

§ 2º Ao determinar a execução do estudo de impacto ambiental e apresentação do RIMA, o órgão estadual competente ou a SEMA ou, quando couber o Município, determinará o prazo para recebimento dos comentários a serem feitos pelos órgãos públicos e demais interessados e, sempre que julgar necessário, promoverá a realização de audiência pública para informação sobre o projeto e seus impactos ambientais e discussão do RIMA.

Art. 12. Esta Resolução entra em vigor na data de sua publicação.

FLÁVIO PEIXOTO DA SILVEIRA - Presidente do Conselho

Este texto não substitui o publicado no DOU, de 17 de fevereiro de 1986.

Annex G - Conama Resolution No. 006 of 16 September 1987

RESOLUÇÃO CONAMA nº 6, de 16 de setembro de 1987
Publicada no DOU, de 22 de outubro de 1987, Seção 1, página 17500

Dispõe sobre o licenciamento ambiental de obras do setor de geração de energia elétrica

O CONSELHO NACIONAL DO MEIO AMBIENTE - CONAMA, no uso de suas atribuições legais,

Considerando a necessidade de que sejam editadas regras gerais para o licenciamento ambiental de obras de grande porte, especialmente aquelas nas quais a União tenha interesse relevante como a geração de energia elétrica, no intuito de harmonizar conceitos e linguagem entre os diversos intervenientes no processo, resolve:

Art. 1º As concessionárias de exploração, geração e distribuição de energia elétrica, ao submeterem seus empreendimentos ao licenciamento ambiental perante o órgão estadual competente, deverão prestar as informações técnicas sobre o mesmo, conforme estabelecem os termos da legislação ambiental e pelos procedimentos definidos nesta Resolução.

Art. 2º Caso o empreendimento necessite ser licenciado por mais de um Estado, pela abrangência de sua área de influência, os órgãos estaduais deverão manter entendimento prévio no sentido de, na medida do possível, uniformizar as exigências.

Parágrafo único. A Secretaria Especial do Meio Ambiente - SEMA[174] supervisionará os entendimentos previstos neste artigo.

Art. 3º Os órgãos estaduais competentes e os demais integrantes do SISNAMA envolvidos no processo de licenciamento, estabelecerão etapas e especificações adequadas às características dos empreendimentos objeto desta Resolução.

Art. 4º Na hipótese dos empreendimentos de aproveitamento hidroelétrico, respeitadas as peculiaridades de cada caso, a Licença Prévia (LP) deverá ser requerida no início do estudo de viabilidade da Usina; a Licença de Instalação (LI) deverá ser obtida antes da realização da Licitação para construção do empreendimento e a Licença de Operação (LO) deverá ser obtida antes do fechamento da barragem.

Art. 5º No caso de usinas termoelétricas, a LP deverá ser requerida no início do estudo de viabilidade; a LI antes do início da efetiva implantação do empreendimento e a LO depois dos testes realizados e antes da efetiva colocação da usina em geração comercial de energia.

Art. 6º No licenciamento de subestações e linhas de transmissão, a LP deve ser requerida no início do planejamento do empreendimento, antes de definida sua localização, ou caminhamento definitivo, a LI, depois de concluído o projeto executivo e antes do início das obras e a LO, antes da entrada em operação comercial.

Art. 7º Os documentos necessários para o licenciamento a que se refere os artigos 4º, 5º e 6º são aqueles discriminados no anexo.

Parágrafo único. Aos órgãos estaduais de meio ambiente licenciadores, caberá solicitar informações complementares, julgadas imprescindíveis ao licenciamento.

---

174 A Secretaria Especial do Meio Ambiente – SEMA, vinculada ao Ministério do Interior, foi extinta pela Lei nº 7.735, de 22 de fevereiro de 1989, que criou o Instituto Brasileiro do Meio Ambiente e dos Recursos Naturais Renováveis – IBAMA. As atribuições em matéria ambiental são atualmente do Ministério do Meio Ambiente.

Art. 8º Caso o empreendimento esteja enquadrado entre as atividades exemplificadas no artigo 2º da Resolução CONAMA nº 1/86, o estudo de impacto ambiental deverá ser encetado, de forma que, quando da solicitação da LP e concessionária tenha condições de apresentar ao(s) órgão(s) estadual(ais) competente(s) um relatório sobre o planejamento dos estudos a serem executados, inclusive cronograma tentativo, de maneira a possibilitar que sejam fixadas as instruções adicionais previstas no parágrafo único do artigo 6º da Resolução CONAMA nº 1/86.

§ 1º As informações constantes de inventário, quando houver, deverão ser transmitidas ao(s) órgão(s) estadual(ais) competente(s) responsável(eis) pelo licenciamento.

§ 2º A emissão da LP somente será feita após a análise e aprovação do RIMA

Art. 9º O estudo de impacto ambiental, a preparação do RIMA, o detalhamento dos aspectos ambientais julgados relevantes a serem desenvolvidos nas várias fases do licenciamento, inclusive o programa de acompanhamento e monitoragem dos impactos, serão acompanhados por técnicos designados para este fim pelo(s) órgão(s) estadual(ais) competente(s).

Art 10. O RIMA deverá ser acessível ao público, na forma do artigo 11 da Resolução CONAMA nº 1/86.

Parágrafo único. O RIMA destinado especificamente ao esclarecimento público das vantagens e conseqüências ambientais do empreendimento deverá ser elaborado de forma a alcançar efetivamente este objetivo, atendido o disposto no parágrafo único do artigo 9º da Resolução CONAMA nº 1/86.

Art. 11. Os demais dados técnicos do estudo de impacto ambiental deverão ser transmitidos ao(s) órgão(s) estadual(ais) competente(s) com a forma e o cronograma estabelecido de acordo com o artigo 8º desta Resolução.

Art. 12. O disposto nesta Resolução será aplicado, considerando-se as etapas de planejamento ou de execução em que se encontra o empreendimento.

§ 1º Caso a etapa prevista para a obtenção da LP ou LI já esteja vencida, a mesma não será expedida.

§ 2º A não expedição da LP ou LI, de acordo com o parágrafo anterior, não dispensa a transmissão aos órgãos estaduais competentes dos estudos ambientais executados por força de necessidade do planejamento e execução do empreendimento.

§ 3º Mesmo vencida a etapa da obtenção da LI, o RIMA deverá ser elaborado segundo as informações disponíveis, além das adicionais que forem requisitadas pelo(s) órgão(s) ambiental(ais) competente(s) para o licenciamento, de maneira a poder tornar públicas as características do empreendimento e suas prováveis conseqüências ambientais e sócio-econômicas.

§ 4º Para o empreendimento que entrou em operação a partir de 1º de fevereiro de 1986, sua regularização se dará pela obtenção da LO, para a qual será necessária a apresentação de RIMA contendo, no mínimo, as seguintes informações: descrição do empreendimento; impactos ambientais positivos e negativos provocados em sua área de influência; descrição das medidas de proteção ambiental e mitigadoras dos impactos ambientais negativos adotados ou em vias de adoção, além de outros estudos ambientais já realizados pela concessionária.

§ 5º Para o empreendimento que entrou em operação anteriormente a 1º de fevereiro de 1986, sua regularização se dará pela obtenção da LO sem a necessidade de apresentação de RIMA, mas com a concessionária encaminhando ao(s) órgão(s) estadual(ais) a descrição geral do empreendimento; a descrição do impacto ambiental provocado e as medidas de proteção adotadas ou em vias de adoção.

Art. 13. Esta Resolução entra em vigor na data de sua publicação.

DENI LINEU SCHWARTZ - Presidente do Conselho

ANNEX

DOCUMENTS REQUIRED FOR LICENSING

LICENCE TYPES	HYDROELECTRIC POWER STATIONS	THERMAL POWER PLANTS	TRANSMISSION NAILS
Preliminary Licence (LP)	• Application for a Prior Licence • MME Order authorising the Feasibility Study • Summary and full Environmental Impact Report (RINLA), when necessary. • Copy of publication of application in LP	• Application for a Prior Licence • Copy of the publication of the LP application • MME Order authorising the Feasibility Study • DNPN research or mining licence, if applicable • City Hall's statement • RIMA (synthetic and integral)	• Application for a Prior Licence • Copy of publication of LP application • RIMA (synthetic and integral)
Installation Licence (LI)	• Feasibility Study Report. • Application for an installation licence. • Copy of the publication granting the LP • Copy of the publication of the LI application • Copy of the Decree granting the concession for the hydroelectric development • Basic Environmental Project	• Application for an Installation Licence • Copy of the publication granting the LP • Copy of the publication of the LI application • Feasibility report approved by DNAEE • Basic Environmental Project	• Application for an Installation Licence • Copy of the publication of the LP concession • Copy of the publication of the LI application • Basic Environmental Project
Operating Licence (LO)	• Application for an Operating Licence • Copy of the Publication of the Granting of the LI • Copy of publication of LO application	• Application for an Operating Licence • Copy of the publication granting the LI • Copy of the publication of the LO application • DNAEE decree approving the Basic Project • MME decree authorising the implementation of the project	• Application for an Operating Licence • Copy of the publication granting the LI • Copy of the publication of the LO application • Copy of the DNAEE Order approving the project • Copy of the MME Ordinance (Administrative Servitude)

This text does not replace the one published in the DOU of 22 October 1987.

MIX
Papier aus verantwortungsvollen Quellen
Paper from responsible sources
FSC® C105338

Printed by Books on Demand GmbH, Norderstedt / Germany